P9-CPV-827

A MILL SHOULD BE BUILD THEREON

An Early History of the Todmorden Mills

ELEANOR DARKE

One of Toronto's most visible landmarks is the towering 1825 brick chimney
bearing the Todmorden Mills name.
Boris Novikoff, East York

NATURAL HERITAGE

Scene at the foot of Pottery Road Hill.
Brick cottage dwelling was the foreman's house and small barn was for the
foreman's use too.Henry Wilson is shown leading the horse,
Thomas John is making hay and Mary Collins is at the water trough.
The large barn and trough are in the paper mill yard.
Painting by artist T. Mower 1838-1934 (known as the Grand Old Man of
Canadian painting).
Courtesy Ann Guthrie, Author, Don Valley Legacy

A MILL SHOULD BE BUILD THEREON

An Early History of the Todmorden Mills

A Mill Should be Build Thereon
Published by Natural History / Natural Heritage Inc.
P.O. Box 95, Station O
Toronto, Ontario M4A 2M8

Copyright © 1995 The East York Historical Society.
The manuscript upon which this work is based is the property of the Borough of East York.
All rights reserved.
No portion of this book, with the exceptions of brief extracts for the purpose of literary review, may be
reproduced in any form without the permission of the publishers.

Design: Derek Chung Tiam Fook
Printed and Bound in Canada by Hignell Printing Limited, Winnipeg, Manitoba
First Printing July 1995

Canadian Cataloguing in Publication Data
Eleanor Darke
    "A Mill Should be Build Thereon: An Early History of the Todmorden Mills," Governor
Simcoe, 1794 : industry and settlement on the Don

ISBN 0-920474-89-6
1. Todmorden Mills (Toronto, Ont.) – History.
2. Toronto (Ont.) – History.   3. Mills and mill-work – Ontario – Toronto – History.
4. Toronto (Ont.) – Industries – History.
5. Land settlement – Ontario – Toronto – History.
I. Title. II. Title: Industry and settlement on the Don.

FC3097.52.D37 1995   971.3'541   C94-930435-2
FI059.5.T686D37 1995

The publisher greatly acknowledges the ongoing support and assistance of the Canada Council, the
Ontario Arts Council, and the Government of Ontario through the Ministry of Culture, Tourism &
Recreation.

The East York Historical Society gratefully acknowledges receipt of a local history publication grant
from the Ontario Heritage Foundation.

Back cover colour separation entitled "Old Paper Mill" Sukimo, oil/board, not dated,
courtesy of the East York Foundation.

CONTENTS

PREFACE

This book has been almost 200 years in the making! In 1795 Upper Canada's (now Ontario's) Lieutenant-Governor, John Graves Simcoe, wrote to Deputy Surveyor General, Christopher Robinson, "...it is necessary that a mill should be build (sic) thereon". In directing the Skinners to build a saw-mill on the Don River near East York's present-day Todmorden Mills Museum, Simcoe began what became, fifty years later, an industrial complex of paper mill, grist mill, brewery, and starch factory, along with the unincorporated village of Todmorden.

Many modern-day people have been involved in getting the story of East York's early beginnings into print. First and foremost, of course, is the author and former curator of the Todmorden Mills Museum, Eleanor Darke, who revisited her manuscript, first written over ten years ago, and prepared it for publication. We are grateful to the Borough of East York for granting us the use of this original manuscript, which Eleanor wrote while she was an employee, as the basis for the book.

We are also grateful to researcher, Ian Wheal, who rechecked the references and assisted in the selection of illustrations. Sincere appreciation is also due to the staff at Todmorden MIlls Heritage Museum and Arts Centre, particularly Curator/Administrator, Susan Hughes, who has kindly provided an epilogue, and Ann Symington, both of whom were most helpful to Eleanor and Ian.

Support funding for the project was graciously provided by the East York Foundation under Chairman Ed Barnett and the Ontario Heritage Foundation. Also, the Council of the Borough of East York and the East York Historical Society gave grants.

Hovering over all of this and making sure all the pieces came together at the right time was our genial publisher, Barry Penhale, whose abiding interest in East York and in heritage publishing were brought together in this endeavor with the happy result you now hold in your hands.

To all of you, our heartfelt thanks.

John S. Ridout, President
East York Historical Society

INTRODUCTION

The research for this book accumulated over many years. Between the tours and meetings, the correspondence and filing, the cataloguing and the phone calls that fill the days of a small museum Curator, I would carefully file each new scrap that appeared and wonder if I would ever be able to assemble it all into a readable history. The temptation to leave everything in files was strong. As long as the material stayed in the research stage, no one could judge my clumsy phrases or disagree with my assumptions. It is very easy to justify an endless collection of research. After all, there will always be more to find. However, with encouragement from John Rempel and others, I screwed up my courage. Here it is, warts and all. I sincerely hope that its readers can find in it some of the fascination I found in unravelling this history of Toronto's earliest industrial village.

Although this book carries my name and I am responsible for any of its flaws, it is not solely my work. First and foremost, appreciation is due to Ian Wheal who did the research on which major portions of it are based. Ian worked as the historical researcher at the Todmorden Mills Museum for two summers under the Summer Canada grant programme of the Department of Manpower and Immigration and maintained his interest in the subject long after the funding ended. For a long time, little sheets of paper containing the most varied snippets of information and sources imaginable kept appearing on my desk marked "from Ian." Ian has since gone on to do marvellous work for a number of other historical agencies in the Province and has acquired an excellent reputation in his field. I was very fortunate to have had his assistance with this project and am grateful to the Federal Government for the grants which permitted his hiring.

Thanks must also be expressed to all the volunteers, historical society members, descendants of local families, libraries and archivists who kept the flow of sources and information coming. They are too numerous to name individually but I hope that they will recognize their work in the finished product and realize the depth of my appreciation.

Eleanor Darke, Toronto

RESEARCHER'S NOTES

What began as a Government of Canada summer research project on the history of the mill-site at Todmorden Mills Museum in May, 1977, quickly escalated by the fall of that year into a voluntary in-depth study based on primary sources in Canada and the United States. I was fortunate that work at East York Public Library (1978) along with another summer contract in 1979 allowed me to stay close to Todmorden Mills and gave me some time to travel to Niagara and Buffalo where major repositories of documents and records existed. The project ultimately consumed a four year period (1977-1981).

Eleanor Darke had made a good start using secondary and other sources from the museum collection including John Ross Robertson's *Landmarks of Toronto* (5 volumes from 1894 to 1914) and Scadding's priceless work, *Toronto of Old* (1873). Unfortunately, a problem of Victorian memory presented itself with the late Victorian generation having only general recollection or unsubstantiated memory of what had taken place some eighty years before. For example William Lea (an 1880's chronicler) and Rufus Skinner (grandson of Timothy Skinner Jr. and a York Pioneer founder) were unable to clearly record events of two generations before their time.

A mirror on the distant past opened, thanks to genealogy-family history, enabling me to reconstruct a skeleton of bare facts. Along with this were excellent early records for other families, which almost made up for the absence of account books, diaries and letter of Todmorden Mills early period 1793-1827.

Using such family history as that of Skinner-Terry, I made Niagara a base of operations and traced back the connection to upper New York State, to New Jersey and other states. Loyalist references such as those at New York City Public Library and at Trenton, New Jersey State Library were invaluable assets.

I soon realized that overall the sources were still fragmentary and selective, and at times I almost leaped for joy when I found a sporadic reference such as one for an 1800 mill petition at Niagara Falls: "Timothy Skinner proposed to build a saw mill near Burch's Mills... 'which are on the river under the bank'... refused." Timothy Skinner (1737-1815) had a long and often unsuccessful quest for a suitable mill-site both before and after he won the Crown Grant to erect mills on the Don River (4 June, 1795).

In his choice of Skinner, Lt. Governor Simcoe showed his knowledge of artisan trades (Skinner was a millwright). Simcoe recognized that American born settlers (such as Skinner) could be of use in the building of Upper Canada. The Crown Grant of Timothy Skinner was drawn up by Simcoe's Deputy Surveyor, Alexander Aitken in York on a date not without significance, for June 4, 1795 was the birthday of King George III.

In searching for records I did not neglect field research and walked the Don Valley many times for clues. I visited distant mill-sites and other centres that may have held a key to the story. In August, 1979 I came across a small road sign marked Skinners Crossing while touring the upper Delaware River Valley of Pennsylvania. The sign was near the spot where Timothy Skinner had built a primitive grist mill over 200 years before I got there.

The research covered a lot of ground. It included an appraisal of what was there today and what may have been there at different time periods. Sources were checked and rechecked and cross-referenced with other sources of other families and sites. It became possible to construct a primitive model of what the site might have contained in a given year, such as 1827.

There were several stages of development of a mill-site, from a small clearing (c.1800), to a larger clearing (c.1810) to a very early factory stage (1827). Likewise mill families usually ran for about three generations before ownership changed or other circumstances intervened. At the Don Mills (Todmorden Mills usual name prior to 1830) Timothy Skinner's sons, Isaiah and Aaron Skinner, were the first millers. Parshall Terry (brother-in-law) and Timothy Jr., a younger son followed. Colin Skinner, son of Isaiah, therefore represented the third generation of millers at the site.

In my research, I found that prior to the War of 1812 the border with the United States was often vague and was no barrier as far as movement of people, technology and trade went. It has been stated elsewhere that "over and over again mills were established by Americans drawing on American experience and American capital."

The wilderness mill whether in Upper Canada or in the United States has a long history of development. None of this is news to mill historians steeped in period research at different historic sites such as Todmorden Mills in East York or at places like Old Sturbridge village in Massachusetts. So the Don River mill-site has a spatial character and a kinship to both its own hinterland and to the areas south of the border.

Research on the John Eastwood and Thomas Helliwell periods

(1821-55) was more difficult due to the lack of substantial sources. William Helliwell's diaries and journals were the best source. They were supplemented by occasional other sources and by newspaper references. Records of the building of William Helliwell's Don Mills plank road, to provide more direct access to Toronto than the earlier Mill Road (Broadview Avenue), and other transportation improvements made it easier to trace the movement of goods and people up and down the Don Valley.

Time did not allow for any extensive research on the Taylor period (1855-1900) at Todmorden Mills, although several good references were found on both Taylor's three mill complex Upper, Middle and Lower Mills, and on the Don Valley Brick Works. There is much yet to research.

Ian Wheal, Toronto

Chapter 1

THE DON VALLEY

Modern-day visitors to the Todmorden Mills Museum experience great difficulty understanding why an important mill village should ever have been built there. The site is squeezed by high speed roads; the mill races and their dam are gone; the hills are covered with scrubby willow and sumach; and the river itself has degenerated into a dirty and nearly dead stream. To understand it is necessary to look past the modern cul-de-sac in which the museum buildings have survived and consider the valley as a whole.

The immediately adjacent portion of the Don Valley is about a quarter of a mile wide and up to a hundred feet deep. It has an interesting and remarkably well-studied geological past. The enormous quarry dug by the Don Valley Brick Company over the years has exposed a complete cross-section of the area's geological layers to the view of experts and has provided us with more information than is usually available.

In pre-glacial times, this valley contained a dense forest and river from which samples of wood, leaves, teeth, bone and shells have been recovered. The varieties of fossilized leaves and shells found indicate that the area enjoyed a climate about 4-5 degrees Fahrenheit higher than at present, approximating that now found in Ohio or Pennsylvania. Among the leaves recovered were those of a form of previously-unknown, now extinct maple. It was named A. torontoniensis in honour of the city.

After the last ice age, the region's average temperature dropped and the water level rose as the glaciers melted and poured their melt waters down the valley to Lake Iroquois. This lake was much larger than present-day Lake Ontario and covered much of today's lower Don Valley. Over centuries, Lake Iroquois slowly drained away leaving behind 250-750 foot deep deposits of cold water clay in the valley. These later provided the raw materials necessary for the valley's brick factories.[1]

Few written descriptions of the Don Valley exist until the late eighteenth century. The valley was explored when Fort Rouillé was established as a fur-trading post (in what are today the grounds of the Canadian

"On the Don," watercolour by George Harlow White, c. 1872. Although it was painted nearly 80 years after Elizabeth Simcoe first visited the Don Valley, this painting portrays the same lushness and beauty that she described.

From the collection of Toronto Historical Board. 1972.19.1

National Exhibition) but the French traders appear to have found the area of little interest. The majority of their trade came down the Humber River so, understandably, their records dealt primarily with the western side of the city.

When the British surveyors and settlers arrived in the late eighteenth century they found that the Don River had two main branches, the East Don and the West Don, as well as a number of tributaries, and that it varied greatly in depth and size. The mouth of the river was sheltered by a peninsula (which later became the Toronto Islands) and had developed a substantial marshy delta. This delta was drained by two main streams which they differentiated as 'the Don' and 'the Little Don.'

In those days, the river drained a very large territory and was much larger and more powerful than it is today. An early resident of the area wrote that:

> *In 1820 when I first came to the Don Mills the roads were*
> *so bad it was the usual custom to raft the lumber from*
> *the tail of the saw mill to the Town of York. At that time,*
> *the volume of water was so large that the rafts were built*

> *... containing from 1500 to 2000 feet ... and a man stand-*
> *ing at one end with a pole in his hand descended the*
> *rapids to dead water, where the raft was left and another*
> *brought down ... to make one large raft of many thou-*
> *sand feet ... As a lad I was very fond of accompanying the*
> *men in this service and have often sailed on rafts from the*
> *Don Mills to York over ground that would scarceley [sic]*
> *float a shingle at the present time.*[2]

The decrease in flow which he records as having occurred by the 1880s was largely caused by agricultural cultivation of the lands which fed the source waters of the river. Our modern flood control dams and storm sewer systems have further contributed to the shrinkage of the river. However, as anyone who has worked beside the river knows, it can still surprise you with its strength. Many an early spring, I've watched the river back up onto the museum grounds and imagined what it must have been like during Hurricane Hazel or one of the many floods recorded by the valley's early settlers.

All early accounts describe the Don Valley as being lush with vegetation. Elizabeth Simcoe, the wife of the province's first Lt.-Governor, delighted in the meadows alongside the river which she said "looked like the meadows of England."[3] In his description of the property given to John Scadding, John Ross Robertson listed some of the valley's vegetation as including "elms of great height and girth, bass-wood, butternut, walnut, wild crab-apples, wild cherry, wild grape, wild currant and gooseberry and prickly ash ..."[4] Dr. Henry Scadding, John Scadding's son, described the view from the Don Bridge at Queen Street as:

> *... very picturesque, especially when the forest which*
> *clothed the banks of the ravine on the right and left, wore*
> *the tints of autumn. Northward, while many fine elms*
> *could be seen towering up from the land on a level with*
> *the water, the bold hills above them and beyond was cov-*
> *ered with lofty pines. Southward in the distance, was a*
> *great stretch of marsh, with the blue lake along the hori-*
> *zon. In the summer this marsh was one vast jungle of tall*
> *flags and reeds ...*[5]

The river originally teemed with fish of all sorts. Robertson wrote of "the river abounding with salmon at the proper seasons, and a number of good fish at all times, rock-bass, perch, pike, eels."[6] J. E. Middleton wrote of the "sea-salmon" coming up the Don ...

> ... in such numbers that in 1795 Abner Miles was selling these splendid fish at 50 cents each. Magrath declared that thirty-pounders were not uncommon. The salmon was usually taken by spearing, after the Indian manner, either in bright sunlight from an overhanging log, or by lighting a fire at night.[7]

Mrs. Simcoe noted that in the winter of 1796 she saw, "Several people ... fishing on the River Don thro' holes cut in the ice; the small red trout they catch are excellent."[8] Even in the 1920s when Middleton was writing, long after the disappearance of the game fish ...

> the long suffering mullet or 'sucker' persisted. To this day schools of 'suckers' crowd the rivers at the time of the Spring freshet and the juvenile of 1923, even as his great-grandfather in the days of his youth, goes fishing with a bushel-and-a-half bag under his arm and scoop shovel over his shoulder.[9]

There is still discussion as to when the salmon actually disappeared from the Don. The date most commonly cited is 1852, when the last salmon is supposed to have been speared at the mouth of the river. However, a later resident, John H. Taylor, reported eating salmon from the Don in the 1860s and the Toronto Field Naturalists were told in 1924 by Mr. C.R. Nash, the Provincial Biologist, that the last salmon had been speared with a pitch fork under the dam at Taylor's Mills, "40 or 45 years ago."[10]

That the area's early settlers could count on an abundance of game for food and furs can be seen in this listing of bounty by John Ross Robertson.

> ... the lands bordering on the stream were alive with genuine game, grouse, quail, woodcock, snipe, plover, sandpiper and wild duck of various denominations; and pigeons innumerable at the proper season; along with numerous fur-producing animals, the mink, the fox, the

*muskrat, the marmot, squirrels in great variety, black,
red, striped and flying, to say nothing of an occasional
deer, bear and wolf.*[11]

Scadding wrote of the "conical huts of the muskrat," the green and
black water snakes, the turtles (snapping and other) and the multitude of
frogs.[12] Elizabeth Simcoe described seeing on different occasions a fine
eagle and "millions of the yellow and black butterflies."[13] Her account also
includes a mention of a much less attractive resident of the valley – the
mosquito.

*July 3, 1796 – Some heavy thunder showers fell this
evening and the mosquitoes more troublesome than ever.
It is scarcely possible to write or use my hands which are
always occupied in killing them or driving them away.*[14]

Many joined her in this dislike since the mosquitoes not only made
life in the valley uncomfortable but, they feared, also carried fevers and
other diseases.

Chapter 2
THE VALLEY'S FIRST INHABITANTS

Records concerning the valley's first inhabitants are scanty and confusing. The Don Valley appears to have been occupied by a variety of First Nation peoples at differing periods. By the time of the first French records for the area, the valley was inhabited by the Hurons. Following the Iroquois attacks of the seventeenth century, the Hurons moved to new areas and joined with other tribes. The Seneca, members of the Iroquois Confederacy, were recorded as being the Toronto area in the 1660s.[15] They later withdrew south of the Great Lakes and were gradually replaced by the Mississauga, part of the Ojibway.

This sketchy chronology is open to easy challenge however. Written records are very sparse and there were few long-term native settlements in the lower Don valley. Even the Huron and the Iroquois moved their agricultural settlements every few years as the soil became exhausted. Their populations were thus in a steady state of movement, adding confusion to the already scanty records.

The Toronto area was a frequent stop on the fur trade route which moved along the Humber River and the French recognized this fact. In 1749 Minister Jonquière wrote the following concerning the decision to establish a fort at Toronto.

> *On being informed that the Indians from the north generally stop at Toronto ... on their way to Chouaguen with their furs we have felt it would be advisable to establish a post at this place and to send there an officer, fifteen soldiers and some workmen to build a small stockaded fort ... Too much care cannot be taken to prevent the said Indians from continuing their trade with the English and to see that they can find at this post all they need as cheap as at Chouaguen.*[16]

Only sparse mentions can be found concerning the native habitation of the Don Valley at the time. Dr. Percy Robinson noted that "another [archaeological] site is on Withrow Avenue east of the Don." According to Helliwell family records, some Mississauga encamped by the Don near their home (at Pottery Road) in 1831[18] and the Taylor family wrote of finding many arrowheads, a stone plow, a tomahawk blade and a flint skinning knife in the valley.[19] Regrettably, the massive highway construction through the valley has destroyed any real hope of the successful archaeological research that would be needed to fill these gaps in our knowledge.

Alexander Aitkens referred to the Don River by the Mississauga name of 'Nechena Qua Kekonk' (appears in various spellings) but never provided any translation of the name.[20] Augustus Jones, another early surveyor, wrote that the native name for it was the 'Wonscateonack' and that this meant "coming from the black, burnt lands."[21] There is a recurrent comment in Toronto histories that the Mississauga used the Don as a route to the Peninsula (now the Toronto Islands) which they regarded as a particularly healthful area.

When Lt. Governor Simcoe desired to establish his temporary capital at Toronto (or York as he named it) he arranged a purchase of the land from the Mississauga. The original treaty of purchase was made in 1787 at the Bay of Quinte and did not specify the exact limits of the land involved. The agreed upon price was £1700 in cash and goods. Another agreement was signed in 1805 to clarify the boundary details and to correct several legal defects in the original purchase.[22]

The extent of the problems with the original agreement can be seen in a number of period sources. Aitken's official survey report of September 15, 1788 mentioned several boundary disputes.

> *I then desired Mr. Lines the Interpreter to Signify to the Indian Chief then on the Spot my intention of beginning to Survey the land purchased from them last year by Sir John Johnson and ... I requested of him to go with me to the spot along with Mr. Lines, which he did but instead of going to the lower end of the Beach which forms the Harbour, he brought me to the River called on the Plan Nichingguakakonk [the Don], which is upwards of three miles nearer the Old Fort than the place you mention in your instructions; he insisted that they had sold the land no further, so that to prevent disputes I had to put it off some days longer ... when Mr. Lines settled with them.*

> ... I continued my Survey Westward until I came to
> Toronto River [the Humber], which the Indians looked
> upon to be the West boundary of the purchase until Col.
> Butler got them prevailed upon to give up to the River
> Tobecoak [Etobicoke Creek] but no further, nor would
> they on any account Suffer me to cross the River with ye
> Boundry line ...[23]

Further problems arose from the distribution of the payments for the
land. Simcoe wrote about this to the Home Secretary, Henry Dundas, say-
ing that:

> The Messissagua [sic] Indians who are the original pro-
> prietors of the land which have been sold to the
> Government make great complaints of not having
> received the presents which they stipulated when they sold
> the lands. Colonel Butler upon my enquiry told me this
> originated from a mistake of Sir John Johnson's who had
> given the Presents to the wrong persons.[24]

No definitive figures exist for the native population of the Toronto
area at the time of British settlement. There are, however, several references to how few there were. Surveyor Joseph Bouchette found that "only
two families of Indians dwelt in the dense forests which then covered the
entire shore"[25] when he visited the Bay of Toronto in 1792.

Simcoe confirmed their sparseness at Toronto in a letter he sent to
Lord Dorchester in 1796 in which he said "... nor is their power to be
slighted since tho' they are not numerous themselves in this part of the
country [emphasis mine], they can draw to a head very formidable num-
bers."[26] This same letter contained one of the few comments of concern
about relations with the native peoples in the Toronto area. In her diary,
Elizabeth Simcoe frequently wrote about meetings with natives. She
recorded receiving gifts of salmon, having them pose for her sketches and
even trusting them to care for her infant son so Lt. Governor Simcoe's fear
was evidently not of violent attack. The fear seems rather to have been that
if the disputes between native and white settlers were not properly
resolved, the Mississauga might chose to close the roads through their territory and thus cut York off from the rest of the province. Chief Justice
Elmsley expressed this concern in a letter he sent to Governor Russell in

1797. "The town of York is nearly 40 miles beyond the most remote of the settlements at the head of the lake and the road to it is through a tract of country in possession of the Mississauga."[27] By 1820, the local native population had decreased so greatly that James Strachan (the brother of John Strachan) could write that "the Indians are no longer a cause of terror to the inhabitants but are disappearing fast."[28]

The security of the road connections to other areas of settlement were not the only areas of concern in York and the Don Valley. The lower Don and many parts of the early town were found to possess conditions unhealthy for their residents. Numerous attacks of 'ague' were blamed on the marshy dampness of

Canaise or Great Sail, an Ojibway Chief who witnessed Toronto's baptism as "York" in August 1793. Drawn by Elizabeth Simcoe in 1794. *Metropolitan Toronto Reference Library T-30535*

the area. There was considerable dissatisfaction with Simcoe's choice for a capital. Russell complained about "the extreme dearness of every building material" and commented that "York is moreover isolated and difficult of access"[29] Indeed, it must have been difficult for the government officials of the time to leave the relative comforts of Niagara, and for some the definite comforts of Quebec, and take up their posts in this completely new and very isolated area.

The job of creating a capital city in a wilderness was a massive one and it is probably not too great a wonder that confusion, mistakes and some injustices occurred alongside the undoubted successes.

One of the successes was the creation of Yonge Street which ran parallel to portions of the Don River. The construction of Yonge not only opened up much more land for settlement, but also resulted in the movement of some of the North-West Company's trade routes to the Don from the Ottawa River. A later source outlined this route as follows: "At York they crossed the Peninsula at the foot of the present Woodbine Avenue, and proceeded up the Don to the Don Mills; from this point they portaged their bateaux on wheels up Yonge Street to Lake Simcoe ..."[30] This decision by the North-West Company likely helped develop the economy of the early Don Mills since their traders took advantage of every opportunity to re-stock their supplies along the route.

One of the early mistakes or injustices affecting the history of the Don settlement was that of John Coon's claim to the mill site property. John Coon is a rather indistinct historical figure. We know that he had been a sergeant in Butler's Rangers,[31] that Mrs. Simcoe visited his farm and home which were about six miles up the Don in 1793 and that he helped the Berczy settlers at Markham in the fall and winter of 1794-5.[32] Clearly he was in early possession of the mill property, yet he was not given title to it. Shortly after Simcoe first arrived in York, the Provincial Secretary, William Jarvis, was asked by Colonel E.B. Littlehales, Lt. Governor Simcoe's aide, to detain John Coon's Deed "till further orders."[33] Coon's petition for ownership eventually was disallowed on the technicality that he didn't know the lot number when applying and the grant was given instead to the Skinner brothers.

If Lot 18 had not contained the best mill site on the river, it is unlikely that Coon would have been denied ownership of the land on which he had been living, regardless of any technical flaws in his petition. Simcoe, however, was very aware of the need for mills in the area and took an active and personal role in the selection of the millers. On the Humber he persuaded other early settlers to move so that King's Saw Mill might be built on their land and John Coon's dispossession seems to have been part of a similar process. If the new capital was to thrive, the desperate shortage of building supplies had to be rectified. Simcoe had to make a choice, and he chose the Skinner family who had experience operating mills both in the United States and in Niagara and who had capital to invest in the property.

John Coon, on the other hand, lacked any milling experience. Based on the quantities of liquor he purchased at Abner Miles' store[34] and the fact that he obtained a still license in 1794,[35] it is assumed that he was running some sort of tavern at the Don Mills, possibly for the North-West Company traders. This may have been the reason that he was later considered qualified to run the government's 'house of entertainment' on the Credit River.[36] This post may have been given to him as an attempt at compensation for the rejection of his land claims. It is interesting to speculate that Simcoe may have had another reason for deciding against Coon. In his *History of Freemasonry* John Ross Robertson says of John Coon's expulsion from the Masonic Order that "Bro. Coon was a profitable customer of the hotel bar" and that "the Brethren ... adjudged the said John

Coon unworthy of ever being admitted into their or any other Lodge."[37]

Given the choice between these two claimants for an economically vital site, it is not hard to understand Simcoe's choice. Just or not, his decision created the first real development of the Don Mills. For an administrator faced with a task the size of that given Simcoe, the ends will always justify the means.

Chapter 3

THE DON MILLS

The government had not originally planned to establish the sawmill for York on the Don. Their first solution to the need for lumber had been to build a government-owned mill on the Humber River and to lease it to a miller. This mill, which was known as the King's Mill, was completed by May 1794 and was leased for five years to John Wilson. It immediately encountered problems. Complaints about the low mill dam, the lack of the specialized saws needed for different woods and the mill's general inability to meet demand abound in early accounts. The 1797 surveyor's report stated that the mill dam was seriously damaged with a large hole in it and that the dam was destroying the Humber fisheries. The fish were unable to get over the mill-dam and some were getting caught up by the mill wheel.[38] These problems were not solved by the next tenant miller either. John McGill wrote to Governor Hunter in 1799 that the miller had "taken sick" leaving the "mill entirely idle" with "further repairs necessary."[39] Fortunately for York, the government had hedged its bets by giving the Skinner family the opportunity to build a privately-owned sawmill on the Don River.

The Skinner family already had a long history in North America, beginning with the emigration of Thomas Skinner from Colchester, England during the reign of Charles I.[40] His grandson, Joseph, was one of the pioneers in the Delaware Valley region of Pennsylvania, exploring it in 1754 and bringing his family there a year or two later. They settled on the west side of the Delaware River just below the mouth of the Callicoon Creek, but Joseph was not fated to enjoy his new property for long. Within a year of settling, he left on a business trip concerning the title to this land and disappeared. His wife eventually accepted his death and returned to Connecticut with her two daughters. Some years later Joseph's remains were found about two miles from his home. He had been shot, but it was never officially determined by whom although unexplained deaths such as his was usually blamed on convenient"unknown Indians."[41]

Joseph's sons remained in the Delaware Valley after his death. Two

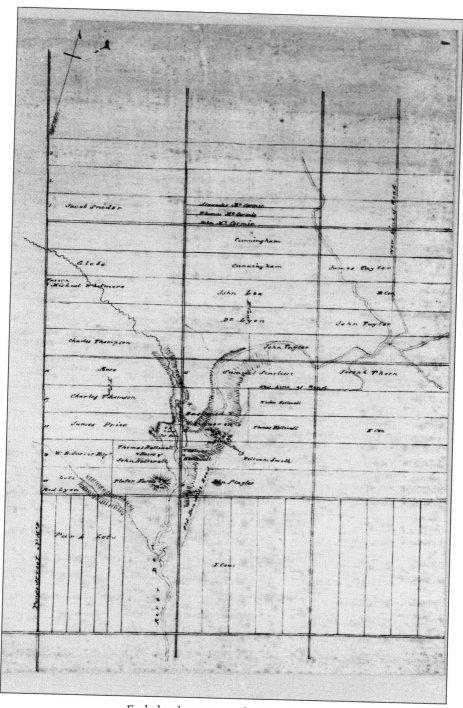

Early land grants on the Don River.
Todmorden Mills Heritage Museum and Arts Centre.

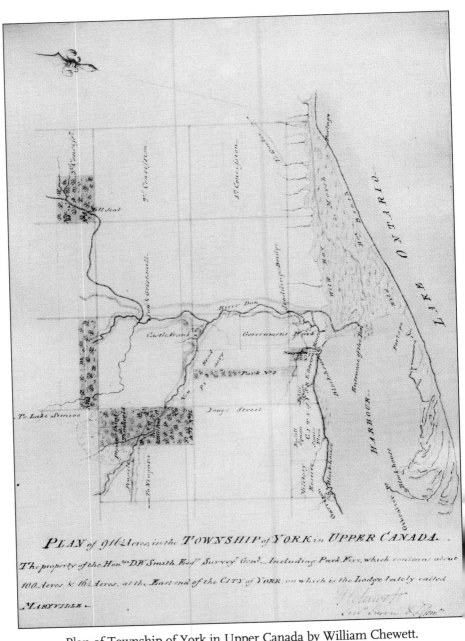

Plan of Township of York in Upper Canada by William Chewett.
Metropolitan Toronto Reference Library S126-B15

of them continued to run the family farm while another, Benjamin, started a new farm near Calkin's Creek with the help of his adult sons. Another son, Timothy, was a millwright and moved around the Delaware Valley helping with the building of the area's first mills.[42] The final brother, Daniel, was also involved in the lumber trade. He became famous in the area for floating the first timber raft down the Delaware, acquiring the nickname of "Lord High Admiral" as a result.[43]

Timothy Skinner moved to Sussex County, New Jersey in the 1760s along with three of his brothers. They acquired substantial properties there, but lost them all during the American Revolution. Timothy and his brothers originally did not support either side in the American Revolution. They tried to do that impossible thing – remain neutral during a civil war. Their homes were in contested territory. They were unfortunate enough to share the name of a prominent British general in the area, Brigadier-General Cortlandt Skinner, and the Revolutionary Governor for the State, William Livingston, was a rabid anti-Loyalist who was described by Lt. Moody as "a man whose conduct had been, in the most abandoned degree, cruel and oppressive to the loyal inhabitants of New Jersey."[44]

Timothy had a large, young family and was unable to flee. The inevitable result came when he was brought before a Revolutionary

"Part of York the Capital of Upper Canada on the Bay of Toronto in Lake Ontario," by Elizabeth Francis Hale, 1804. The Skinner sawmill supplied lumber for many of the first houses in York.
National Archives of Canada C-34334

Committee and convicted of being a Loyalist for refusing to swear an oath of allegiance to the new regime. Timothy described his sentence in a later appeal he made to the Crown for compensation for his losses.

> *That your Memorialist in the time of the late unhappy disturbances in America, was settled in Sussex County in the Province of New Jersey but being thought inimical to the Laws of the United States was imprisoned for the space of 14 Months and fined in the sum of £168 New Jersey Currency.*[46]

His claim for compensation was disallowed because "The Claimant not having come within the British Lines during the War the Commissioners do not consider him as a Loyalist."[47]

The repressive actions of the Revolutionary Committees had the usual effect of forcing people to take sides. Some of Timothy's family took the oath. Others fled. His elder brother, Benjamin, joined the British at Oswego in June 1778. The 1811 land grant petition of his son, Ebenezer, includes the information that when Benjamin got to Oswego he worked (as a civilian) for Joseph Brant and was "Used by Capt. Caldwell and a number of men belonging to Butler's Rangers ... with others employed in foraging parties till in the month of August following."[48]

Timothy was finally able to sell his properties in New Jersey in 1783 and emigrated to Canada via Sorel, Quebec as part of a group of "Civilian Loyalists."[49] He and his family moved to Niagara in July 1784 and settled on Portage Road, Stamford Township, near the Falls. Over the next fifteen years, he and his brothers received the maximum allowances of free land for their categories. Timothy went back to his old trade of millwright and farmed near Niagara Falls. He may also have operated a flour mill for the North-West Company. An estimate of the quantities of land granted to the family can been obtained by considering that granted to Timothy alone. He received 100 acres in 1793, 200 in 1796, 100 in 1797 and a final 890 in 1798. The Don Mills formed only a small percent of the family's acreage, but held the greatest potential for profit since most of their other lands were undeveloped and scattered throughout the province.

The grant of land in the Don Valley was not made to Timothy, but rather to his two oldest sons, Isaiah and Aaron Skinner. The process began after 1793 when they received permission to buy Lot 19, Concession 2 from its original nominee, Abraham Laraway. At the same time Simcoe gave them Lot 18 which lay immediately north of Lot 19, on the condition that they erect a sawmill there at their own expense. While no document exists stating this condition, it is confirmed by Simcoe's letter to Christoper Robinson in 1795 in which he stated that "it is necessary that a mill should be build thereon. It is therefore I directed [the] Skinners to build a mill."[50] The Skinners built their mill early in 1795 and petitioned to have the grant confirmed in June 1795, the requirements having been completed.[51]

Problems about the location of this mill quickly arose when it was discovered that it had not been built on Lot 19 or even on Lot 18, their other property, but rather on Lot 13. The surveying of the area was still incomplete at this time and there is an unlikely possibility that the Skinner thought Lot 18 crossed the river when it is Lot 13 which does so. Simcoe's directive to them clearly stated, however, that they were to "erect Mills on Lot Number Eighteen on the West side of the River Don ..." [emphasis mine][52] Attempts to rectify the situation only led to more

Castle Frank, the Simcoe's summer home on the Don River, by Elizabeth Simcoe, 1796. Lumber for this house was purchased from the Skinner sawmill in 1796-7.
Archives of Ontario, F 47-11-1-0-228

confusion. The Minutes of the Council meeting at which they were grant-
ed Lot 13 in 1796 incorrectly transcribed their request as follows,
"Petitioners having erected a Mill on Lot 12 pray for Lot 13 adjoining on the
East side of the River Don. Granted."[53] The Skinners never owned Lot 12.
Lot 12 is not even on the Don River.

However it was done, the mess finally was straightened out and
the sawmill became very busy supplying lumber for many of the first
houses in York. In 1796-7, the Skinners supplied some of the lumber
used to construct Simcoe's own home, Castle Frank, for which they
were paid £9, 5s, 2d.[54] Construction of Castle Frank had begun in late
1794.

Their success with the sawmill was rewarded in July 1796 when
Simcoe gave them permission to erect a grist mill. He also issued the
following instruction to the Commissar of Stores, John McGill.

> *York 14th July 1796*
> *Sir*
> *You are hereby required and directed to give out of His
> Majesty's Stores in your charge to Isaiah Skinner one
> pair of mill stones and a complete set of grist mill irons,
> as an encouragement for him to build a mill on the Don
> for the accommodation of the new settlements in the
> neighbourhood of York, and for so doing this shall be
> your order and authority.[55]*

The Skinner grist mill was built and in full operation by the summer of 1797. Describing the mills, an early resident of the area, William
Lea wrote that:

> *These mills were of great importance, being the only
> ones near York. The grist mill had only one run of
> stones and was kept going day and night as was also the
> saw mill. The people brought their wheat as far as from
> Hamilton and many other ports on the lake. The grain
> was taken up the Don in boats to the Sugar Loaf Hill
> and thence up the flats by ox teams to the mill. People
> living at a distance with no roads were in the habit of
> taking a bushel of wheat in a bag on their backs follow-
> ing the trail in the woods to the mill.[56]*

Skinner's Mill. Painted by Elizabeth Simcoe in 1796. Unfortunately for historians, she found the trees more interesting subjects than the mill itself.
Archvies of Ontario, F-47-11-1-0-229

While these mills encountered many of the same problems as the sawmill on the Humber, the on-site presence of their owners kept them in better condition. One recurrent problem was the requirement that the mill dam not obstruct the migration of the fish. This condition was written into all mill permits for both the Don and Humber Rivers but, as we saw earlier, failed in its intent to preserve the salmon and other game fish.

While the Skinner mills would not seem very grand to our modern eyes, they filled a real need in their own time. The only known illustration of them was done by Mrs. Simcoe and does not indicate a particularly imposing structure. The sawmill was most likely little more than a one-storey, largely open wooden shed. William Helliwell, writing thirty years later, described the property as "consisting of a grist mill and a saw mill of the most primitive description."[57] Sawmills of the same period in the United States were generally small open-sided sheds with an outside, undershot water wheel. Grist mills were also small, but two-storey in height and somewhat more enclosed. Both types of mills and their machinery were built almost entirely of wood. The saws were of the upright frame type, described at the time as going "up today and down tomorrow."

The undershot style of wheel is driven by the rush of water under it and is much less efficient than an overshot wheel. Since it required a much smaller mill dam and almost no mill pond, however, it was easier to build and consequently favoured in new areas.[58] The Skinner's mill dam most likely was built of loose stones, trees and underbrush and had a fall of no more than eight feet.

The miller's dues were set by the Legislature at a twelfth of the grain ground and wood sawed.[59] Being a miller may have been profitable, but it was also risky and difficult. There was the danger to the mill itself of fire from the friction in the wooden cogs in summer and from flood every spring. The miller risked death or maiming if he fell or caught his clothing in the large gears. He also faced the hazard of having to crawl around the water wheel in the mill-race to chip away ice to keep the machinery running as late in the year as possible.[60]

It is difficult to determine what standard of life was maintained at the Don Mills during these years. The Skinners had most likely built themselves small log or frame houses similar in size and layout to the Scadding cabin or to the kitchen portion of the Terry House in the Todmorden Mills Museum. We know that they bought many of their supplies at Abner Miles' store on Queen Street. His 'Day Books' of accounts show purchases by them of meat, whiskey, leather, an almanac and shoes. They also show that the Skinners bought a stove from him in 1796 for £3, a sum which they paid off in installments. This must have been one of the earliest stoves in the area, since most homes relied on open fires for both heating and cooking until after the first quarter of the next century. A stove would have made their home considerably warmer and more comfortable than those of their neighbours.

Another feature which may be inferred from Abner Miles' records is the existence of at least one brick chimney. Miles' accounts show that Isaiah Skinner paid Hugh Monteith £4 on May 11, 1796.[61] Hugh Monteith is listed in early census reports as a brickmaker and it was probably his brickyard which the Playter Diary referred to in April 1804 as the "old brickyard" at the beach.[62] This brickyard appears to have been built near the foot of modern-day Parliament Street and to have been the first brickyard in the lower Don Valley. In addition to selling lumber and flour the Skinners gathered barrels of salmon and fruits in the valley and traded them at Miles' store.

The Skinners' neighbours in the valley were of varied backgrounds, ranging from criminals to members of the nobility. In 1798 and 1799 there were at least two depositions sworn out against local residents for the crimes of break and enter and assault. The first, in 1798, was laid by Isaiah Skinner against John Welch, charging that "John Welch did break down the door of the said Isaiah Skinner and forcibly entered the house."[63] The second and more serious charge was laid against Samuel Sinclair in 1799 by Charles Falcondo. It is intriguing how clearly the atmosphere of the whole sordid situation comes through both the legal language and the years.

> *Deposition of Charles Falcondo against Samuel Sinclair, York*
>
> *... deponent was ... violently assaulted, beat, battered and bruised contrary to His Majesty's Peace by Samuel Sinclair, Mariner, with a Spade or Shovel and called a Black Son of a Bitch, for no other reason than for giving Mr. Ruggles an order on the said Sam'l Sinclair for Eight Shillings which he owed to this informant, this Informant prays that the said Samuel Sinclair may be dealt with according to law and further saith not.*
>
> > *His*
> > *Charles X Falcondo*
> > *Mark*[64]

Both cases were settled out of court and all charges dropped.

The Skinners had some more prominent and law-abiding neighbours. Their most famous neighbour was, of course, Lt. Governor Simcoe, whose summer home, Castle Frank, was built just south of them on the west side of the Don River. This house was a relatively simple building in the design of a 'Greek temple', an architectural concept created by its columned porch of peeled tree trunks.

Another famous neighbour was John Scadding, who came to the area as Simcoe's estate manager. His home was built in the 1790s at the corner

"A View of Scadding's Bridge and House on the Don River 1794,"
by Sir Edmund Wyly Grier after a drawing by Elizabeth Simcoe.
Produced for John Ross Robertson in 1896.
Metropolitan Toronto Reference Library, John Ross Robertson Collection, T-11488

of what are today Gerrard Street and Broadview Avenue. He had been granted the whole of Lot 15, Concession I which consisted of 180 acres running from the bay north to modern-day Danforth Avenue, and bounded on the east by Broadview Avenue and on the west by the Don River.[65]

In those days, the Scadding farm was well-known because of its prominent location near the Don Bridge. Some early maps even call it "Scadding's Bridge."[66] This house was saved from destruction in the late nineteenth century by the York Pioneers Historical Society, whose members moved it to the Exhibition Grounds and opened it as a museum. It remains there today and is the oldest house in Metropolitan Toronto. It is still operated as a museum by the York Pioneers. John Scadding's son, Henry, wrote one of the earliest histories of Toronto, thus providing an invaluable resource to later researchers.

The Playter family, who lived south of the Don Mills, were another interesting group of Skinner neighbours. Records of the New Jersey Historical Society describe the head of this family, George Playter, as:

> *an English cabinet maker, settled at ... Burlington County, New Jersey, where his property, valued at £1000 in currency, was confiscated and sold. He joined the British at Trenton in 1776 and was employed in repairing a bridge, which "proved of great service." Later he was employed in obtaining intelligence for the army and continued on service until the Peace, when he joined his family in Pennsylvania.*[67]

George Playter originally came from Suffolk, England where he had married a Quaker and joined that denomination. Despite the pacifist principles of his religion, his military services to the Crown were of sufficient value to entitle him to a life-time pension as well as the usual land grants.

Accompanied by his large family, he moved to York in May 1795 and completed construction of a home on his grant of valley land by 1796. Mrs. Simcoe visited there that year and described the bridge that the Playters had built across the Don River.

Wed., July 6th, 1796
I passed Playter's picturesque bridge over the Don; it is a butternut tree fallen across the river, the branches still growing full leaf. Mrs. Playter being timorous, a pole was fastened through the branches to hold on by. Having attempted to pass it, I was determined to proceed but was

Playter's Bridge, drawn by Elizabeth Simcoe. Mrs. Simcoe visited the Playter family in 1796 and recorded her attempt to cross "Playter's picturesque bridge... a butternut tree fallen across the river, the branches still growing full leaf."
Archvies of Ontario, F-47-11-1-0-234

frightened before I got half way.[68]

Evidence that the Playter family retained their Quaker beliefs after the Revolution can be seen in the pronoun forms they used in correspondence such as this note to John McGill concerning payment for their labours in the construction of Castle Frank.

March 15th, 1797
Respected Friend

Sergeant Lyndan informs me thou are desireous I shall Sled some boards from Skinner's Mills, to Castle Frank. If I do, thou must pay me one quarter of a Dollar for every hundred feet; not exceeding one inch thick; which I have, or may Sled, as it is the customary price, and really it is worth it.

I am thyn to Serve
&c
George Playter[69]

One of George Playter's sons, Eli (born 1775), became a member of the Assembly for York County in 1824. He was a moderate Reformer and supported a number of Reform bills but, as a result of his disputes with the Family Compact, found it advisable to return to United States in 1827.

Although the mill land grants had been made to both Isaiah and Aaron Skinner, Aaron did not remain in the area for long, selling his share to Isaiah and returning to Niagara around June 1797. His name ceases to appear in Abner Miles' accounts at that time while that of his brother-in-law, Parshall Terry, first appears the following month. Isaiah is the only Skinner named on the July 1797 list of inhabitants of the area. This list demonstrates clearly how sparsely the area was still settled with only 59 people listed as living in the entire Don Valley area south of the Forks (near the present Leaside Bridge.)

Inhabitants of the Don and the Marsh

Head of Household	Male	Female
George Playter	4	3
Isaiah Skinner	3	2
Parshall Terry	4	4
George Porter	2	2
Jacob Phillips	2	2
Jacob Winters	1	1
John Playter	1	2
Frederick Brown	2	2
William Cornell	7	3
John Ashbridge	3	1
Benjamin Morley	1	1
Patrick Burns	2	1

Single Men
Hugh Monteith
Sam'l St. Clair [problably Sinclair]
Sam'l Whipple[70]

In that same year, the Town of York had only slightly more than 200 residents.[71]

In 1797, Isaiah Skinner petitioned for even more land in the area and for a town lot in York. Both requests were refused. The following year, however, the area did receive a major improvement – a road to the mills. It was called the Don Mills Road and was laid along the top of the east side of the valley from the Queen Street Bridge to the mills, thus following approximately the same route as modern-day Broadview Avenue. William Lea later described this road as "a mere wagon track, winding among the trees and underwood without grading or any improvement save the cutting of the underbrush and the removing of the fallen trees."[72]

This road had two major drawbacks. It entered the city at the Queen Street Bridge for which there was a toll charge and it went down to the mills via an extremely steep hill which became virtually impassable in bad weather. Despite these disadvantages it remained the area's main road until 1841.

In 1796, Isaiah's brother-in-law, Parshall Terry, was granted Lot 7, Concession II which lay immediately south-east of the mills. In 1799 he bought a one-third share in the mills, half of the east part of Lot 18 and one-third of its west part from Isaiah for £500. Presumably this sum also paid for the lease which he received at this same time to operate the mills.

Isaiah moved back to Niagara shortly afterwards, although he retained a two-third share in the property until 1801 when he transferred it to his father, Timothy Skinner Sr. In Niagara he went into partnership with Oliver Grace, a former Royal Navy officer, and became captain of the sloop *Mary Ann*. This sailing vessel was one of the earliest merchant vessels to regularly serve York.[73] Isaiah was styled Captain Skinner or Skipper Skinner in various references and operated the *Mary Ann* for three or four years. Isaiah vanishes from Niagara after 1806 and is thought to have returned to the United States shortly before the War of 1812. He was still alive in 1810 when his father, Timothy Sr., mentioned him in a reference. He was probably deceased by 1815, the year of his father's death since he is not mentioned in his will. In 1806 he sold his share in the *Mary Ann* to his partner, Oliver Grace. When he returned to the States, he left his son, Colin, behind in Niagara with Timothy Sr., the boy's grandfather. Colin was then at the age of apprenticeship and, judging by his later career, had probably begun his training as a millwright.

Oliver Grace, Isaiah's old sailing partner, moved to the United States in 1812 to escape the problems caused by his son, Oliver Grace Jr., who was charged with high treason for joining the American troops during their raids into Canada during that year. Many disaffected Loyalist settlers, including several members of the Skinner family, returned to the United States around 1812 as, once again, the family was split by political allegiances they would have preferred to ignore.

In 1800, Timothy Sr. sought permission to build mills on his land in Niagara. Although his request was rejected, the specifications for these mills were filed with the request and are worth considering in any attempt to picture the size of those at the Don Mills. His proposed saw-mill was to have been 26' x 12' and the grist mill 28' square. Both were to be built within three years of permission being granted.

With Simcoe's departure from the province in the summer of 1796, the Skinners seem to have lost some of their government contacts. Also, their in-law Terry was no longer a member of the Legislative Assembly and they may finally have pushed their demands to the limit. Whatever the reason, after 1798, they found increasing difficulty in winning approval for their petitions and tenders. Not only did they lose out on the Niagara mills request, but they were rejected as lessees for the King's Saw Mill on the Humber. Their tender for this lease had included an offer to build a new sawmill and a grist mill with two sets of stones at their own expense (estimated at £1500 -£2000 New York currency.)[74]

Another possibility is, that as land became more valuable and the criteria for Loyalist standing became more rigid, Timothy's war-time attempt at neutrality began to hurt their status. In 1805 his daughter, Sarah, requested land as the child of a United Empire Loyalist and Timothy's status was officially investigated. Her claim was disallowed on the ground that her father did not "join the Royal Standard or reside within the British Lines previous to the Treaty of Separation in 1783" and that it was therefore recommended by the Committee "that the name of Timothy Skinner be expunged from the U.E.[L] List."[75] This re-assessment of their status was not unique. Between 1800 and 1810 many American-born settlers were removed from the lists of United Empire Loyalists – a grievance which led some to support the American forces during the War of 1812.

Indeed the government had begun to sound somewhat testy about all these Skinner land requests. In 1818, Timothy Sr.'s brother, Haggai, petitioned for an additional 200 acres and was rejected with the notation that "The petitioner acknowledges to have received 600 acres of land – he ought to be satisfied with the indulgence he has met with."[76]

Chapter 4

PARSHALL TERRY

Although Parshall Terry owned a share in the Don Mills for only six years and spent most of that time trying to sell them, it is his name which is most commonly attached to them. Time after time, we read comments in the history books such as "On the east bank of the Don ... a road known as the Mill Road led to Parshall Terry's Mills."[77]

Like the Skinners, Terry came from a family with a long North American history. The founder of the family in America was Richard Terry (1618-76), one of whose great-grandsons, Johnathan Terry, married

Butler's Rangers, in which Parshall Terry served as Lieutenant, operated out of Fort Niagara. Fort Niagara, shown here in a 1784 watercolour, was the strongest of the western posts.
Old Fort Niagara Association

Jemima Parshall. They named their first son Parshall Terry and (perhaps just to confuse future historians and readers) the same name was given to both his son and grandson.

The first Parshall (1734-1811?) was one of the earliest settlers of the Connecticut Susquehannah Purchase and appears in the early records of that settlement as a 'Proprietor.' Terrytown, Pennsylvania is named for the family. The history of the Susquehannah Purchase is both confusing and bloody. Both Connecticut and Pennsylvania claimed ownership of the territory and the native inhabitants resisted the settlement claiming, with considerable justification, that those from whom the purchase had been made lacked the authority to make such a sale.

Nathaniel Terry, the brother of Parshall (I), was killed in an attack in 1763. According to family legend, Parshall saved himself by playing dead.[78] On another occasion, a writ was issued for his arrest as one of the participants in a riot which occurred at Fort Wyoming in January 1771, during which a man was killed.[79]

Parshall (I) had at least three sons, one of whom was also named Parshall. Parshall (II) was born about 1754 and grew up in the frontier community of the Susquehannah Purchase. He made a number of fur-trading and trapping expeditions into Canadian territory and the Ohio Valley during the years before the American Revolution. These were, of course, illegal since access to those areas were restricted to the Montreal trading companies and the resident native tribes.

When the American Revolution began in 1776, Parshall and his father both joined the Revolutionary Army. Parshall (I) served as a private in the 24th Regiment of the Connecticut Militia under Zebulon Butler, while Parshall (II) joined the First Westmoreland Company.

Parshall (II) didn't remain in the Revolutionary Army for long. One family history says that while on a march he dropped out to tie his shoe-laces and was insulted and struck by an officer. He returned the blow and deserted on January 11, 1777 rather than face the stiff punishment likely to be meted out for such an offense. Whether or not this story gives a true reason for his desertion, it is documented that by June 15, 1777 he had switched sides in the conflict, appearing on the payroll of the Indian Department of the British Army with the rank of Lieutenant in Butler's Rangers at a daily rate of 4 shillings New York currency.[80] Butler's Rangers were an early irregular unit who fought alongside the Six Nations Iroquois. Terry served with the Rangers for the rest of the war, taking part in their attack on his old home territory in 1778. According to family tradition, he met members of his own family among the survivors of that attack who disowned him, cursing him as a traitor. The rumours arising from this family rift must have been particularly pernicious. Many years later, his son felt it necessary to submit documents of his grandfather's death to the newspapers of the area as proof that the old man had long out-lived the Revolution and had not been murdered by his son during the "Wyoming Massacre."[81]

In 1775 or 1776 Parshall Terry (II) married Melia (Amelia?) Stevens (1758-1789). Their first child, Parshall (III), was born in 1776. Melia accompanied her husband when he deserted, bearing another child, Mary, in 1780 and a third, Martha, in 1783.[82] They were recorded on the "Roll of Loyalists at and In the Dependencies of Cataraqui and the Provisions They Draw" in December 1783. Terry and his family moved to Niagara the fol-lowing spring where he took up land in the Township of Willoughby, near the mouth of the Black Creek. His name appears on a survey done of that area in July 1784.[83]

Little is recorded about Parshall and Melia's life during those first few years of settlement. Most of the Loyalists at Niagara found it a difficult time. The winter of 1788-89 was recorded as being exceptionally bad due to the previous summer's crop failure. Melia and Parshall had two more children during this time, a girl named Submission, and a boy, William. Melia died in 1789.

Around the time of her death, Terry began his climb up the local gov-ernment ladder. He became deputy-commissary under John Warren and was appointed a magistrate for the Fort Erie area in 1789.[84] In 1792 he was elected to the First Parliament of Upper Canada as the representative for the Fourth Riding of Lincoln and Norfolk.[85] He does not appear to have been a particularly conscientious member, being more often absent than present, but this post gave him some social prominence and likely accounts for the tradition of attributing the Don Mills to him.

In 1793 or 1794 Terry married Rhoda Skinner, who had been born in 1775 and was thus twenty-one years his junior. Rhoda was one of Timothy Skinner's daughters, a sister of Isaiah and Aaron. Rhoda and Terry began a second family, having seven more children over the next eleven years; five girls and two boys.

At the completion of his term as a member of the Legislature in 1796, Terry obtained the post of magistrate for the Home (York) District. He held this post for the remainder of his life.[86] It is also about this time that he seems to have begun seeking, and obtaining, large quantities of land. Possibly his new in-laws taught him how easy it was to do! He began by gaining title for the land he'd been living on since 1784. By 1803, he had added almost 2,000 more acres to his holdings.

May 9, 1797	*- 500 acres (Bertie Township)*
May 9, 1797	*- 700 acres (Willoughby Township)*
April 16, 1799	*- 100 acres (Bertie Township)*
September 4, 1801	*- 200 acres (Burford Township)*
November 4, 1803	*- 225 acres (Scarborough Township)*

He also bought the 200 acres of Lot 7, Concession II near the Don Mills, a share in the mills in 1799, part of the land at the Forks of the Don, and 200 acres from Samuel Sinclair in Vaughan Township. He also arranged to have 200 acres in Markham Township granted to his brother Nathan, then bought it from him in 1807.[87]

Nathan Terry came to Canada in 1800 and lived with Parshall for almost a year while applying for land. It is evident that whatever hard feelings might have existed in the family during the Revolution had died down considerably over the intervening years. Nathan returned to the United States in 1807 after selling his land to Parshall. Terry also applied for and received 200 acres on behalf of his wife, Rhoda, as the daughter of a U.E.L. His eldest son, Parshall (III) applied for and received 200 acres of land near Newmarket in 1800. Other purchases may also be found, since Terry speculated with his lands, buying and selling continually.

By July 1797, Terry had moved his family to the Mill area where he had constructed a small frame house. Although questioned by a number of historical architects, tradition claims that this house now forms the kitchen portion of a house which is still standing on the Todmorden Museum grounds. It would have been an extremely small home, especially considering it housed eight people in 1797, but if it were indeed the house they lived in it had the conveniences of a real brick fireplace, a number of glass windows, a solid wood floor and a basement with a well.

In 1799 Terry built another sawmill further up the river on his property at the Forks of the Don and immediately began advertising it for sale.

SAWMILL FOR SALE

FOR SALE - One half of that excellent mill known by the name of the "New Sawmill" situate upon the River Don,

> ... together with one hundred acres of land thereunto
> belonging. For particulars apply to Parshall Terry.[88]

Terry seems to have spent at least as much time trying to sell his properties as actually working them. In 1800 he also began advertising the Don Mills for sale in the *Upper Canada Gazette*.

> One-third share of that valuable Grist and Saw-Mill on the River Don known by the name of Skinner's Mills and also a lease on the other two-thirds, from the first day of May last, for eleven years to come. The above Mills within two miles and a half of the town of York, on a living stream, where the Purchaser may grind, saw or Raft Boards in the dry and sell thereon. There are four hundred acres of land, belonging to the said Mills, on which is plenty of good pine. For further particulars, enquire of the subscriber on the premises.[89]

Terry continued to run both of these ads with a variety of options and descriptions until 1804.

He also seems to have tried his hand at shingle-making at the Mills and at a variety of cash-crops. When the Government advertised their need for hemp for ships' roping in 1802, he planted hempseed. In 1803 when a need for linen was advertised, he planted flax. He also either grew large quantities of potatoes, turnips and strawberries or received them in payment from his mill customers.[90]

Terry maintained his involvement in local politics, acting as Pound-Keeper for the area in 1799 and as "Overseer of Ways [Roads]" in 1800.[91] He was frequently away from home on business during these early years and relied on his son, Parshall, and sons-in-law, Joseph Lutz (husband of Mary) and Alexander Galloway (husband of Submission) to manage the mills. By 1805 all of the children from his first marriage had obtained land of their own and had moved away from the mills and Terry had to begin hiring local residents such as Samuel Sinclair and the Kendrick brothers.

Terry began to experience financial difficulties in 1804 when a severe fall flood put all of his mills on the Don out of service for about three months right at harvest time. The extent of this flood was described in Eli Playter's diary entries for the period of September 11-17, 1804. "The fences gone, all the stacks of hay and grain ... Everything that could float did ...

Silhouette of Alexander Wood. Wood was the local agent for Sir Richard Cartwright who held mortgages on Terry's Don Valley properties from 1805-1808.
Metropolitan Toronto Reference Library, York Pioneer and Historical Society, 971.354.Y.59, p. 26

We thought more of saving our hay than the Sabbath." The flood was followed in October by a heavy snow which made repairs difficult. During the repairs, Terry strengthened his existing mills and added a grist mill to his property at the Forks of the Don.[92]

Either because of these unexpected costs and loss of income or because of his failure to sell some of his more speculative properties, Terry was forced to sell his share of the mills back to Isaiah Skinner and to take a mortgage on his other valley property from Sir Richard Cartwright of Kingston in 1805. He began to default on this mortgage after the first year. Scadding recorded a drop in the profits of millers as beginning in 1804 and continuing to 1812 which may provide an explanation for his problems in repaying the loan. The reports of Cartwright's agent in York contain comments like the following:

> *22 April 1807 – Mr. Terry has not paid me any money but says he will deliver flour, will it suit you?*
> *9 June 1807 – Mr. Terry promises to convert his flour into money if possible.*
> *26 May 1808 –- Mr. Terry some days ago sent a letter*

here for you ... probably this may be some new proposal.
I shall tell him what you say on the subject of the lands
of his you hold by mortgage when I see him.[93]

In June 1808, Terry sold Cartwright the east half of Lot 13 and 200 acres of Lot 5, Concession 3 for the sum of £150 and the debt was resolved. Cartwright retained ownership of the east half of Lot 13 until 1832 but there is no record of any member of the Cartwright family ever living there.[94]

Just as they rid themselves of financial problems, tragedy struck the family. Parshall Terry drowned in the Don on his way back to the mills from York. The following notice of his death appeared in the *York Gazette* on July 23, 1808.

DROWNED IN THE DON

Departed this life on the 20th, Mr. Parshall Terry. His death was occasioned by his getting into the River Don on horseback. By this misfortune an exemplary wife and large, helpless family are left to the care of the all-disposing Providence, and a resistless appeal is made to the benevolence and sympathetic generosity of a virtuous public. The particular situation of the road near the Don bridge, calls imperiously upon the commissioners appointed by his Excellency for the particular care of the roads and employing the voted money for immediate repairs, as many lives are seriously threatened with danger by its present state, in consequence of the causeway being removed by an excessive flood. The place, when seen, suggests the nature of the required improvement, and as a part of duty we earnestly recommend it to public attention.[95]

In his *Landmarks of Toronto*, John Ross Robertson adds the information that Terry had "essayed to ford the Don on horseback ... some fifty yards north of the present Queen Street bridge. He was swept away, his body being afterwards found near the mouth of the river, but his horse reached the shore."[96]

Rhoda now faced a terrible situation. Her oldest child, Aimie, was only thirteen. Her step-children had their own families and would have found it very difficult to support a step-mother and her seven children. Her father was elderly and likely unwell, having made his will the previous year.[97] Rhoda was not educated to operate a business and thus could not have run the mills even had she been able to afford to hire sufficient labourers.

She did, however, have one option – the most common one available to women of her time. On May 18, 1809, ten months after the death of her husband, she married William Cornell. His wife, Content Davis, had died five months before in childbirth. Some speed was justified. Rhoda had seven children, aged 4-14 years, to support and Cornell had twelve children, aged 2-20 years, who needed a mother.

As if raising nineteen children weren't sufficient, Rhoda had six more children by Cornell – Edward (1810), Susannah (1812), Harriet (1814), Charles (1816), Charlotte (1819) and Frances (1821).[98] Even in those days, this size of family occasioned some comment. Charles Fothergill stopped at the Cornell's in 1817 on his "Journey from Montreal to Upper Canada," and wrote:

> *4 March 1817 – William Cornell, who is a 'ci-devant'*
> *Quaker doing well with his farms and mills. A house full*
> *of fine girls. He was a widower and married a widow for*
> *his second wife. Between them have 23 children living,*
> *have had 28. Five daughters are married.*[99]

Rhoda's final two children were not yet born when this was written. It is not clear to whom the five deceased children referred to by Fothergill had been born.

William Cornell (1766-1860) was born in Rhode Island and served with the British during the American Revolution in Colonel Carleton's Legions. After the Revolution he brought his family across the lake in his own boat where they lived until he had received his permanent land grant. He was given a small property on the Don in 1797 and a larger one in Scarborough shortly afterwards, where he built his home.

Cornell was operating a lucrative trading business with his ship between Oswego, New York and York when he married Rhoda. Both this ship and its cargo of wheat were seized at Oswego as a war prize in 1812. He also built Scarborough's first grist and sawmills and set out the Township's first orchard in 1802.[100]

Rhoda surrendered Terry's lease of the mill property property to her father, Timothy Sr., before she re-married. Timothy Sr. sent his younger son, Timothy Jr., to York to operate the mills while he attempted to sell them, but without success.

Rhoda, who acquired the nickname 'Tiny' within the Cornell family, bore her last child in 1821 at the age of 46. She died thirteen years later. Cornell was then 68 and did not re-marry. He died in 1860 aged 93 years and 6 months.[101] Rhoda and Parshall's daughters married comfortably. One, Aimie, married Isaac Cornell, a son of William Cornell by his first wife. Her son, Timothy, joined the Mormon faith as an adult along with his half-brother Parshall (III) and eventually moved to Utah.

Chapter 5

A SKINNER RETURNS

With the marriage of Rhoda Terry to William Cornell in 1809, management of the Don Mills was assumed by Rhoda's brother, Timothy Skinner, Jr. Timothy Skinner Jr., then aged 42, had been married to Ann Lutz since 1800. They had three children: Joseph (1801-76), Timothy (1802-1812) and Mary Ann (1803-1885.) In 1811, Timothy Sr. bought Isaiah's third share of the mill property, thus acquiring complete ownership. He then gave a two-third share to Timothy Jr. The upper saw and grist mills that had been built by Terry at the Forks of the Don were sold to David Secord, a relative of Laura Secord.

At the outbreak of war in 1812, Timothy Jr. joined the Lincoln Militia, along with many of the other men in his family. The 1812 military census of the Lincoln Militia listed four Skinners – Joel, Job, Colin and Haggai – as artillerymen (out of only seven men so classified.) Both Timothy Jr. and Haggai Sr. were also listed under the category of "Over Age."[102] Five other members of the Skinner family served in other units. Despite his age classification, Timothy Jr. was on active duty with the unit until his death.

Timothy Jr. was granted absence from his unit to tend to his family and mills in York on at least two occasions in 1813 and 1814 but, while he was away fighting, the mills were managed by Samuel Sinclair. Sinclair had been wounded during Brock's successful attack on Detroit in 1812 and invalided home.[103] He had settled in the Don Valley about 1796 and owned land north of the mills. Early records list his occupation as "mariner,"[104] but he appears to have left that trade entirely by 1800.

The Don Valley was raided by American troops during the War of 1812. Although within the framework of the whole war these forays were insignificant, they were no doubt occasions of great concern and excitement to the area's inhabitants. The first visit was to the Playter properties in 1813. The Playter sons were all officers in the militia and the Americans hoped to capture them in their raid. They failed in this, although they did succeed in capturing their elderly father, George Playter. He gave his 'parole' not to fight in that war and was released.

According to Robertson, the Playter property was also targeted because "many of the archives of the Province were conveyed to their residences for safety, but that precaution was in vain for the invaders found out where they had been placed and carried away all they could lay their hands on."[105] According to Eli Playter, they didn't just take the government records. They also stole his sword, razor, jewellery and some clothing.[106]

His diary goes on to delight in the fact that the American soldiers hadn't succeeded in getting all that they were looking for in the valley. The Playter sons, with the help of some of their neighbours, including Samuel Sinclair, had undertaken to remove two boatloads of ammunition from the Garrison in York before it was abandoned. These boats were brought across the Bay and up the Don to the north end of the Playter property where the ammunition was buried and the boats scuttled. The first boat made it up the river without trouble, but the second became stuck at the 'Big Bend' (near today's Riverdale Farm) and had to be partially unloaded before it could be re-floated. The Americans were reportedly in hot pursuit of these boats, having been informed of them by 'traitors' in York.

Timothy Skinner Jr. enlisted in the Lincoln Militia at the outbreak of war in 1812 along with several other members of the Skinner family. The Lincoln Militia took part in the Battle of Queenston Heights, shown here in a painting by James B. Dennis who also fought there. Timothy Skinner was listed as missing in action after the Battle of Chippewa in 1814 and later declared dead.
Samuel E. Weir Collection and Library of Art, Queenston, Ontario.

Fortunately, the pursuing Americans were unfamiliar with the river and gave up the chase when they also got stuck at the Bend.[107]

It was lucky for the Don Mills that the American troops failed to advance any farther up the river. Mills were prime 'industrial' targets and, with their owner an active enemy combatant, it is likely that they would have been destroyed. There was also considerable excitement at the mouth of the Don. A large war frigate was under construction there. The retreating British and Canadian troops burned it and all its stores to prevent them from falling into American hands.

Timothy Jr. was listed as missing in action after the Battle of Chippewa on July 5, 1814.[108] Although it was assumed he was dead, his wife could not sell his property or re-marry until she could find witnesses prepared to swear that they saw him die. Such legal proceedings were, of course, delayed until the end of the war when the witnesses returned from service.

Sinclair leased the mills from Timothy Sr. and Ann Skinner in 1814 and operated them along with George Casner, a Pennsylvania German who had been farming further up the valley since 1796. Ann Skinner moved into York with her children where she was able to send them to school for the first time. The school they attended was kept by Mrs. Glennon, a widow, in her home at the corner of George and Duchess [Adelaide] Streets. This house consisted of one large room which was partitioned by a curtain into a school and living quarters. Mrs. Glennon had a large number of pupils of all ages and, although she was an educated woman, she had no training as a teacher. All of these factors combined to make the maintenance of discipline in her classroom extremely difficult. One of the worst offenders was her own daughter, Theresa. According to the memoirs of the Reverend John Carroll who attended the school as a boy, Theresa was particularly unruly one day. Mary Ann Skinner, "a stout, strong young woman," offered to help Mrs. Glennon punish her. Theresa didn't care for this idea and a battle ensued with "Miss Skinner getting a punch in the eye which left it painful and inflamed." Honours, however, were about even since Theresa got "a good thrashing."[109]

Over the next three years, Samuel Sinclair appears to have kept an eye on the mills for Ann and an equally close eye on Ann for himself. They were married once Timothy was legally declared dead. This marriage probably took place in 1817, since Ann's daughter, Mary Ann, is listed as Mary Ann Sinclair in that year.

Samuel Sinclair remained in charge of the Mills until November

1820 and was assisted by George Casner throughout this time. Sinclair purchased Parshall Terry's mills at the Fork of the Don from David Secord in 1817 and ran them until 1821 when he sold them to D'Arcy Boulton.

Ann's marriage to Samuel Sinclair seems to have been happily uneventful. As far as can be determined she had no more children. Sinclair continued to assist with the neighbourhood mills on occasion, but his main efforts went into farming his land on the Don. He acquired some local significance from his gift of land to the 'Primitive Methodist Connexion' in 1851 for the construction of a church and graveyard. He had been holding this land in trust for this purpose for several years and records of tent church meetings on the site appear in the early 1840s.[110] This is now the site of the Don Mills United Church. Their graveyard, together with that of the Taylor family, are the oldest ones remaining in the Borough of East York. Samuel Sinclair died in 1852 at the age of eighty-five and is buried there beside Ann.

The local economy was slow immediately after the end of the War of 1812, but by 1820 it was bustling. A great variety of new mills were erected along the Don and its tributary streams. These new mills were naturally more 'modern' and larger and for the first time the Don Mills began to

Don Mills Primitive Methodist Church. The Taylor family funded the construction of this church on land held in trust for it for many years by Samuel Sinclair.
Metropolitan Toronto Reference Library

suffer from competition. Although Sinclair had made improvements in and about the mill and its dam they were not sufficient to bring them up to the improved standards of the new mills.

Both of Ann's children were underage at the time of their father's death. Sinclair was appointed their guardian and, with Ann's approval, sold their shares in the Don Mills to their uncle Henry Skinner in 1817. Although Skinner family tradition holds that Timothy's son, Joseph, had second thoughts about this when he came of age in 1822, he took no legal action to dispute the sale until thirty years later.

Mary Ann married John Hayes in 1823 and moved to his property in Whitchurch Township, remaining there until her death in 1885. Joseph married Sarah Anderson in 1824. After working for a time as a blacksmith in Whitchurch Township, York County, he opened a general store on Yonge Street which he operated until his death in 1876. His son, Rufus, also became a merchant in York. Rufus acquired a captain's commission and served as a 'Quarter master' during the Fenian Raids in 1866. Active in the Orange Lodge, he was also one of the founding members of the York Pioneers Historical Society.

A complete changing of the guard occurred at the Don Mills in 1820. Timothy Skinner Sr. had died in 1815 and left a share in the mills to his grandson, Colin.[III] In 1820 Colin purchased the remaining shares of the property from his uncle, Henry Skinner, and moved there. He immediately sold two-thirds of the property to two recent English immigrants, Thomas Helliwell and John Eastwood. These three new owners came to the Don Mills with new capital, and more importantly, with new ideas. They were to completely change the area over the next thirty years.

Chapter 6

THE DON VALLEY

Between 1820 and 1850 the town of York became the city of Toronto and its population increased thirty-fold.[112] Similarly, the population of the Don Valley increased and the area acquired many new industries. A tannery was built a little north of the Don Mills; Bloor's Brewery began operations on Castle Frank Brook; and Alexander Milne built a sawmill and later a large woollen, fulling and carding mill on the eastern branch of the Don River.

Change also came to the Don Mills. The Helliwell family (of whom more will be said in the next chapter) constructed a brewery and new grist mill and Colin Skinner went into partnership with John Eastwood to construct the first paper mill in the valley.

Colin Skinner (1792-1841) was the son of Isaiah Skinner. He had remained behind in Niagara serving an apprenticeship as a millwright when his father returned to the United States in 1806. When Timothy Skinner Sr., Colin's grandfather died, his will stated (in part) that he did "give and bequeath to my Grand son Colin Skinner one hundred acres of the west end of lot no. 14 in the County of York, Township of York and second concession, to him and the male heirs of his body lawfully begotten for ever ..."[113] In 1817, Colin's uncle, Henry Skinner, had bought a two-thirds share of the Don Mills from his brother's widow, Ann, and in 1820 sold it to Colin.

Colin immediately sold most of the land to John Eastwood, retaining only the mills and about ten acres. John Eastwood had gone into part-

John Eastwood,
from *Paper in the Making.*

nership with his father-in-law, Thomas Helliwell, to make this purchase. At the same time Helliwell and Eastwood bought the east half of Lot 14. In 1821 the Eastwood and Helliwell families moved to the valley to join Colin Skinner. Shortly after they arrived, they advertised the establishment of a "tin manufactury [for] all types of tinware."[114] John Eastwood had been trained as a tin-smith, plumber and glazier in England, as had one of Thomas Helliwell's sons, John.

The Helliwell Brewery (called the Don Brewery in most accounts) began operations in 1821 and it is likely that Eastwood had a hand in its construction. His tin-smithing skills would certainly have come in handy for the construction of the vats and piping. For unknown reasons, the partnership between Eastwood and Helliwell ended in 1822 and the property was divided between them. Helliwell retained the land in Lot 14 and the brewery and the small part of land in Lot 13 on which it and his house stood. Eastwood, the junior partner, received the balance of the land in Lot 13. He immediately began a new partnership with Colin Skinner and together they added a distillery to the old saw and grist mills.

In 1823 Thomas Helliwell bought a strip of land from them on Lot 13 consisting of "one acre, one rood and eighteen perches," in order to obtain room to build a raceway off the existing mill race.[115] This shared millrace was to prove a source of contention between the two groups for years afterwards.

Colin Skinner had been trained as a millwright and had served in the Lincoln Militia throughout the War of 1812. He purchased property in the Township of Yarmouth, Western District (London area) shortly after 1812 and probably lived there until he moved to the Don Mills. In 1824 Colin

---

J. EASTWOOD, & CO.,
PAPER MANUFACTURERS.
OFFICE AND AGENCY,
IN THE STORE OF MR. JOHN BENTLEY,
No. 65 Yonge-St., Toronto.

Advertisement for the J. Eastwood & Co. Paper Mill, 1851.
Colin Skinner died in 1841 and Eastwood bought his share of the business.
Toronto Directory, 1850-51, York County, p. 198, Metropolitan Toronto Reference Library

Skinner married Thomas Helliwell's daughter, Mary. They had one child, Colin Jr.

John Eastwood (1788-1850) was born to a middle-class family which lived near Todmorden in Yorkshire, England. He was listed in an 1811 town directory there as "John Eastwood, gentleman."[116] In 1816, he married Elizabeth Helliwell (1792-1864), the eldest daughter of Thomas Helliwell, and they emigrated to Canada via New York the following year. In 1818 they were joined by his wife's family, who settled in Drummondville (Lundy's Lane and Queenston Road) where they ran a general store. In 1818 they moved to Niagara Falls where they leased and operated a distillery.

Eastwood was very active in the politics of York. He was elected as an alderman for the St. Lawrence Ward of the city in 1836 and again for the St. Andrew Ward in 1841. He was a member of the inaugural Council for the Township of York in 1850 and was elected by that Council to the post of Deputy-Reeve, but died before completing his term of office. In politics he was considered a Reformer and was an early supporter of William Lyon Mackenzie, but he was unwilling to continue this support once treason was contemplated. Robertson describes the resulting, almost-comic situation this way.

> Both Mr. Eastwood and Mr. Skinner were in sympathy with Mr. Mackenzie's political views at this time and continued to be until they saw that they were leading to open rebellion. Shortly before the actual outbreak of the rebellion it is related that Mr. Mackenzie went up to the mills to persuade their proprietors to join his forces. Mr. Eastwood was not at home and Mr. Skinner when he saw Mr. Mackenzie coming, ran and hid himself, so that he could not be found.[117]

Eastwood remained under suspicion by the government for several years after the 1837 Rebellion both because of his political support for Mackenzie and because of a promissory note for £75 they found from Mackenzie to him. It was dated September 29, 1837 and the government suggested that it had been a contribution to Mackenzie's military plans. Eastwood asserted that it was simply a business debt owed by Mackenzie for the purchase of newspaper. Some accounts state that Mackenzie paid the debt to Eastwood's heirs during the 1850s, after his return from exile.

Eastwood was active in the Masonic Lodge.[118] Some accounts credit him with founding the St. Andrew's Lodge in Toronto. He was certainly one its first members and his name appears in many of its records.[119] He was also one of the founding members of the Unitarian Congregation in Toronto.

In one of the first issues of his newspaper, *The Colonial Advocate*, William Lyon Mackenzie pointed out that printing paper was coming in duty free from the United States and that the government should enable a papermaker to establish a mill in Upper Canada. He claimed that this would save them £3,000 a year and held a public meeting on the subject in 1825 in the Masonic Rooms in York, where a petition was prepared and sent to the Assembly asking them to provide a financial incentive for such a mill. The Assembly agreed and in 1826 offered a premium of £125 to the first person to build a producing paper mill in the province.[120]

A race then began between James Crooks of West Flamboro and Eastwood and Skinner of the Don Mills, seeking to be the first to produce paper and win the award. Eastwood and Skinner added onto the old grist mill building in order to convert it into a paper mill and advertised for a tin pedlar to collect rags. They also made the following announcement in *The Colonial Advocate* concerning their intentions.

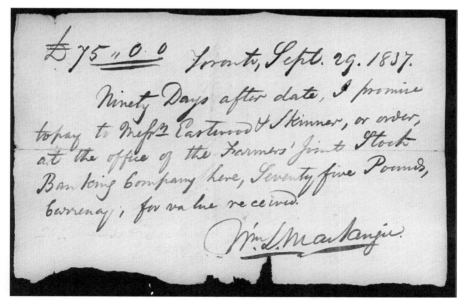

Promissory note from William Lyon Mackenzie to John Eastwood.
Both Eastwood and Skinner were Reform sympathizers before the Rebellion.
Metropolitan Toronto Reference Library

YORK PAPER MILL

The subscribers, having entered into copartnership, for
the purpose of converting the Don Mills into a paper mill,
and having most of the millwright work done, and also a
considerable stock of rags collected, they will lose no time
in completing and carrying it into operation.
One of the partners, having had many years experience
in the business, both in England and the United States,
they flatter themselves that they will be able to make a
good article, which they intend to sell as cheap as it can
be had from the State of New York.
They earnestly solicit the patronage of the public and
hope all persons will feel an interest in saving the RAGS,
as they have persons appointed to collect the same, who
will call regularly and give the highest prices for them.

JOHN EASTWOOD
COLIN SKINNER
York, 17th April, 1826   ROBERT STONEHOUSE[121]

Robert Stonehouse never appears again on any list of partners, so his

WANTED.

TWO Boys 15 or 16 years of age as apprentices to the Paper Making business, to whom liberal encouragement will be given, by application to

EASTWOOD & SKINNER.

York Paper Mill,
May 26th, 1831.   15-tf

Newspaper advertisement for apprentices in March 3, 1832 issue of *The Courier.*
Metropolitan Toronto Reference Library, Vol. 3, No. 16, page 4

connection with the mill must have been very short. It is also likely that he was the experienced paper-maker of whom they bragged. Neither Eastwood nor Skinner appear to have had any previous experience in this field.

The Don paper mill produced its first paper in August 1827, but lost the award to James Crooks' Flamboro mill. It is somewhat amazing that almost every local account provides a different reason for Crooks' success. William Lea claimed in 1881 that Crooks "desiring not to be beaten, gathered a quantity of old paper and by boiling and stirring with a paddle, formed a pulp which he made into paper."[122] Nearly 50 years later, another local historian, Charles Sauriol, wrote that "Crooks was able to beat Eastwood and Skinner ... only by starting his mill on a Sunday."[123] In reality, the race was never that close. Crooks produced his first acceptable paper a good eight months before Eastwood and Skinner were able to do so. It seems safer to believe the 1829 official government report which stated that Crooks had won because he "already had the frame of a grist mill built when the act was passed which he was able to convert into a paper mill."[124]

Mackenzie visited the Don mill shortly after it began production and described the area in his *Sketches of Canada*.

> *About three miles out of town in the bottom of a deep ravine, watered by the River Don and bounded also by beautiful and verdant flats are situated the York Paper Mills, distillery and grist mill of Messrs. Eastwood & Co; also Mr. Shepard's axe-grinding machinery and Messrs. Helliwells' large and extensive brewery. I went out to view these improvements and returned much gratified with witnessing the paper manufacture in active production.[125]*

In the fall of 1828, Eastwood and Skinner advertised for a three-year £300 loan in order to make further improvements and offered as security "a paper and grist mill, a saw mill, a distillery and 200 acres of land free of all encumbrances."[126] The loan was obtained and they began to purchase a quantity of paper-making machinery from the United States, thus becoming the owners of the first machine-made paper mill in the province. In 1829 they petitioned the government for the return of the duties they had already and would still need to pay. The Assembly approved the remittance of these fees on the understanding that the partners would have to prove, in future, that they had made an effort to purchase the items from Britain.[127]

Eastwood and Skinner's Reformist sympathies may have hurt their business prospects with the government. Certainly they had not made any sales to them prior to 1828 when Mackenzie wrote about the situation in his usual emphatic manner.

> *As a proof of the disposition for discouraging home man-*
> *ufacture displayed by the Maitland Government, we*
> *state that on enquiring at Mr. Eastwood, we learn that*
> *the two Houses of Parliament, and all the offices of*
> *Government, have never purchased nor used of York*
> *paper, in all more than ONE solitary REAM. Deeds*
> *show!*[128]

The *Journals* of the House of Assembly were being printed on paper from Crooks' mill at this time, so it is probable that he was receiving the majority of the government business.

Eastwood and Skinner acquired Mackenzie as a customer sometime around August 1830. He was definitely using their paper for the printing of *The Colonial Advocate* in 1834. In the February 24th issue of that year he stated that "We have again to apologize for the bad quality of paper furnished by Messrs. Eastwood & Skinner. Our readers may rely on it that as soon as we can remedy the evil we will not fail to do so, and the remedy shall be a permanent one." The problem he complained about was a type of creasing caused by the new mechanical dryers. The problem was indeed remedied since it does not appear in later issues of the paper.[129]

The switch to machine-made paper and the opportunities afforded the mill by its closeness to York quickly made the paper mill extremely profitable. Both partners were soon able to afford to live most comfortably.

It is supposed that when he first came to the mills, Colin Skinner moved into the house attributed to Parshall Terry. It is uncertain when its front four rooms were added. If Colin was responsible for their construction, they were probably built around the time of his marriage in 1824. By the 1830s, maps of the area show that a large frame home had been constructed on the north side of Pottery Road, directly opposite the Terry house. Traditionally, this was called the Skinner house so was probably the home he built from his profits in the mill.

Where Eastwood lived when he first arrived at the mills is uncertain. Robertson's sketch of the mills shows a small house alongside the paper mill which might have served this purpose. Eastwood's second home was built in 1832 and was a much grander structure. Built of stone, it had two stories and was surprisingly asymmetrical. While at first glance it appears purely Georgian in design, a closer look reveals it to have three front windows and an off-centre front door. Although the house was destroyed by fire in July 1926, one of its later residents described its appearance when she lived there as a child. Miss Ash's father was a manager for the Taylor paper mills at the turn of the century and rented the Eastwood House from them. The Taylors had acquired it in 1855 along with the mills. Among the features which she remembered was the ballroom which she described as being "large enough for my brothers to circle about on a bicycle – which they did in winter."[130] This ballroom took up the front half of the house on the two-window side. Compared to the houses which had preceded it at the Don Mills, it was a very elaborate home indeed. Such a large house probably required servants and the Eastwood's were likely the first mill-

Eastwood House. Watercolour by Owen Staples, 1912.
Metropolitan Toronto Reference Library, John Ross Robertson Collection, T-11207

owners to employ such a staff at the Don Mills.

The period of 'roughing it' and early settlement seems to have ended in the Don Valley by this time. As the following 1832 advertisement for George Playter's property indicates, it was now considered a comfortable country district, conveniently located near to the city.

FARM GEORGE PLAYTER

> *1 1/2 miles from York lying on the River Don. To be sold late George Playter, Esquire farm and containing 117 acres, 40 of which is clear and intervale with a good orchard of apple, pear, Plumb and Chery and other fruit trees and a fine stream of water running through it abounding with trout and other fish. The Farm is well adapted for a dairy or a Gentleman's country seat.*
> *WATSON PLAYTER*　　*Whitchurch*
> *JOHN PLAYTER*　　*York*
> *GEORGE PLAYTER*　　*Newmarket*[131]

Phrases such as 'city advantages' must always be understood within

It is believed that this house was that of Colin Skinner. Shown here in 1945 converted to a restaurant. Note that the papermill had also been converted for use as a riding stable.
Penhale Collection

The Taylor paper mills, ca. 1877, from *Toronto Illustrated Past & Present*.
The Taylor family bought the Eastwood and Skinner mill from the Eastwood
estate in 1855. It is shown here as "The Lower Mill."
Metropolitan Toronto Reference Library

the context of their time. York in 1832 was not what we would regard as a city. As the following population statistics show, it was growing incredibly rapidly although the actual figures themselves seem incredibly small by modern standards.

1815 - 720	1831 - 3,969
1828 - 2,235	1832 - 5,505
1829 - 2,511	1833 - 6,094
1830 - 2,860	1834 - 9,252[132]

These statistics also make the following anecdote, which is attributed to Charles Lord Helliwell, fully believable.

> In 1831 a man, driving a load of pork down Yonge street, was followed by a bear. It was in the evening and the man did not observe the bear. He arrived at the tavern [Bosworth's Tavern on the north-east corner of King and Yonge Streets], unhooked his horses and went to bed in the tavern. In the night the inmates were aroused, and found that the bear had gotten into the horses' stall and was creating quite a sensation.[133]

Mackenzie's printing business was seized and sold at auction after the Rebellion. Eastwood is said to have purchased at least one of Mackenzie's printing presses there and to have used it to establish his own printing business.[134] He began with the production of a monthly agricultural journal called the *Canadian Farmer and Mechanic*. Around 1846 he either re-named this journal or replaced it with one called *The Cultivator*, which continued in publication until late in the nineteenth century. He also published *The Toronto Almanac and Canada Calendar for 1842*, in addition to other almanacs. His printing shop was located in Toronto at Yonge and King Streets. By 1846 this firm was described as "paper and blank book manufacturers, stationers and school book publishers." The business also operated a retail store in Hamilton, run by Eastwood's second son Daniel, where they sold "a full and varied assortment of Fancy Stationery, American cheap publications, etc."[135] By 1850 they had opened another stationery store, on King Street in Toronto which was managed by his wife's nephews, Colin Skinner Jr. and Charles Lord Helliwell.

Colin Skinner Sr. died in 1841 and his son (who survived him by only ten years) sold his father's share of the mill business to John Eastwood. Eastwood died November 17, 1850. One of his sons, Colin Skinner Eastwood, then aged twenty-six and acting as his father's executor, operated the paper mill until 1852. At that time Joseph Skinner pressed his long-deferred claim on the property stating that it should not have been sold while he was a minor. Records concerning the mill over the next three years are confusing. Joseph Skinner appears to have been at the mill-site during part of this time. According to Skinner family tradition he even operated the mill for a period, but knew nothing about making paper so was unsuccessful. By 1855 the Eastwood family had reached some sort of agreement with Joseph Skinner and sold the mills and their property to John Taylor and Brothers. The Eastwood sons and Colin Skinner Jr. continued operating the Eastwood publishing business.

The Taylor family had moved to the Don Valley in the 1830s. In 1846 they had established a competing paper mill further up the Don River from the Don Mills. With their acquisition in 1855 of the Eastwood and Skinner mill they began their domination of the valley's economy, a domination which lasted into the twentieth century.

By this time, an identifiable village of mill-owners, mill-workers, shops and shop-keepers had taken root in the valley and along the Don

View of the Todmorden valley, Yorkshire. John Eastwood is credited with naming the village which grew around the Don Mills 'Todmorden' in memory of his original Yorkshire home. The similarities in the landscape are striking.
Todmorden Mills Heritage Museum and Arts Centre.

Mills Road. The village had even acquired its own school – a small, one-room structure which stood near the modern-day intersection of O'Connor Drive and Donlands Avenue.[136] Eastwood is credited with naming the village 'Todmorden' after his (and the Helliwell family's) original home in Yorkshire.[137] It is commonly stated that the area acquired this name in the 1830s, but it did not appear in print until 1851 when it is found in W.H. Smith's book, *Canada, Past, Present and Future.*

> ... *a mile and a half from the City, the road crosses the River and about a mile further on you reach the village of Todmorden. There are but a few houses on the upper bank, but on descending by a steep and circuitous road to the Valley below, you reach a paper mill, a grist mill and starch factory; with the residences of the owners and work people employed.*[138]

Chapter 7

"A VIGOROUS AND SUBSTANTIAL FAMILY"

The Helliwell family moved to the Don Valley and began to make their mark on the Don Mills at the same time as Colin Skinner and John Eastwood. John Scadding mentions them in the following passage:

> In 1821, and down to 1849, the Mill Road was regarded chiefly as an approach to the multifarious works ... founded near the site of Parshall Terry's Mills, by the Helliwells, a vigorous and substantial Yorkshire family, whose heads first settled and commenced operations on the brink of Niagara Falls, on the Canadian side in 1818, but then in 1821 transferred themselves to the upper valley of the Don, where that river becomes a shallow, rapid stream, and where the surroundings are on a small scale, quite Alpine in character ...[139]

The Helliwell family came from the area around Todmorden, England, as did John Eastwood. Todmorden was then in Yorkshire, right on the border with Lancashire. An article about the town quoted a doubtless apocryphal story of "the lady in the border town of Todmorden [who] uncertain of where she lived, asked at the Post Office and when told that she was in Yorkshire replied, "Thank God for that! I believe they have terrible weather in Lancashire."[140] In recent years, however, Lancashire has gotten the last laugh as the re-distribution of county boundaries has moved Todmorden into Lancashire.

More is known about the Helliwells than about any of the other Don mill owners since William Helliwell kept a diary throughout his life. Although not all of his diaries have survived, those that remain provide a clearer picture of their years at the Don Mills than we can ever hope to obtain for any other period of its existence. In addition, one of William's sons summarized some of the diaries into a family biography using information from some of the diaries which have since disappeared.

This biography includes the following passage, in which William Helliwell describes his family's English background.

> My father was a small cotton manufacturer and also occupied and owned a small farm. At that time there were no power looms the manufacturer only spinning the cotton and the weavers taking the yarn home and wove it into calico by hand looms. Every small stream that afforded a power sufficient to drive a spinning and carding machine were utilized. This was before the advent of steam. After carding and spinning, the yarn was given out to weavers who took it home and wove it into calico and returned it to the 'Cotton Masters' who took it to Manchester or Halifax to large capitalists who bleached and printed it and fitted it for exportation.[141]

The Helliwell family had first established their small textile cottage industry in Wickenberry Clough near Todmorden in the early 1700s. They also farmed a property known as Houghstone Farm. Later that century, they added the Ratcher (or Ratcha) property to their possessions and used it to produce what they described as "stuff goods."[142] Although this mill was larger than anything they had owned previously, it was still not large enough to remain economically viable in the rapidly changing cotton industry of the early 1800s.

In 1817 the Helliwell family sold the Houghstone farm to John Fielden for approximately £400 and used this money to follow their son-in-law, John Eastwood, to Canada. John Fielden later became famous as one of the cotton lords of England. Thomas Helliwell Sr. sailed from Sunderland to Quebec in April 1817 and the rest of the family left in June from Liverpool bound for New York. William explained this peculiar method of emigration when he wrote "The reason for my father taking a different route and at a different time from his family was that skilled manufacturers were not allowed to leave the country and he had to do it clandestinely and precede them in order to prepare a home for them."[143]

Thomas' family consisted of his wife, Sarah Lord; five sons; Thomas, John, Joseph, William and Charles; and two daughters, Elizabeth and Mary. His eldest daughter, Elizabeth, was already settled in Canada with her husband, John Eastwood. Sarah Lord and her children had a long and difficult voyage to New York and were only able to join Elizabeth and her husband in Niagara early in 1818. Thomas Sr. arrived even later, having stopped in Montreal to purchase the goods with which to open a general store. They opened this store in early summer 1818 at the corner of Lundy's Lane and the Queenston Road (now the site of Drummondville.)

The following year, in 1819, the family separated again as Thomas, Sarah Lord and some of the younger children moved to the Niagara Falls where they rented a distillery from Christoper Bughmer. William remained behind to attend school at Drummondville Hill. During the next two years, the Helliwells bought their share of the Don Mills property.

The family moved to the

Sarah Lord Helliwell (1773-1842). Her son, William, described her as "a splendid business woman."
Todmorden Mills Heritage Museum and Arts Centre

Don Valley in stages. Thomas Sr. and the older sons came first and built the brewery, a malt house, a distillery and a small home for the family. The rest of the family joined them in 1822. The brewery was immediately successful, selling its produce both to taverns and to the public. Thomas Jr., who had married while living in Niagara, operated the brewery's retail business from his store in the Toronto market square. His brothers John and Joseph, who were running a tin-smithing business in Toronto, lived with his family in their house which stood on the west side of market square. Their younger brother William also stayed in this house while he attended the York School on Colborne Street.

Later in the nineteenth century, John Ross Robertson used the information given him by long-term residents of the area to sketch and describe the Helliwell properties on the Don. The brewery he described as being "a building of two stories, about fifty feet square, constructed of stone, brick and wood ... Connected with the brewery and in the same building was a distillery." He further wrote that "close by the side of the brewery was originally a small frame house built by Thomas Helliwell, Sr. This was torn down [and] a stone dwelling put up on its site."[144]

Thomas Helliwell Sr. died in 1823 and left the business to his wife and sons. By the conditions of his will, each son was to be taken on as a partner at the age of 21. Elizabeth and Mary each received cash bequests. Mary, who had recently become engaged to Colin Skinner, also received "One yoke of steers, one cow and calf, my feather bed and bedding, six chairs and one Table and other small furniture, paid out of the Profits of the Business."[145] As William recorded, the business "was carried on by my Mother ... who was a splendid business woman and my three older brothers, Thomas, John and Joseph." The eldest brother, Thomas Helliwell Jr., was married twice. He had six children by his first wife, Mary Wilson, who died in early 1833, probably of cancer. William's diary recorded about six months before her death that a "canser" [sic] had been removed from her breast that day and that she "was as well as could be expected."[146] Thomas married Ann Ashworth later that year, with whom he had seven more children. Thomas Jr. was the senior partner in the family business, handling all the administrative work from his office in town as well as managing their beer store. He was also involved in the financial community of York and was elected a Director of the Bank of Upper Canada by its shareholders in 1834.

William and Jane Helliwell and daughters.
Todmorden Mills Heritage Museum and Arts Centre

John Helliwell died in 1828, leaving a wife and two children. With his death, the family gave up their tin-smithing business in York and concentrated their activities on the brewery and other Don properties.

The third Helliwell son, Joseph, was married three times: first to Sarah Glassco, who bore him nine children; secondly to Harriet Round, whom he met on a business trip to Britain; and thirdly to another Englishwoman, whose name is not recorded. Harriet had at least three children and it is thought that the third wife had none. Less is known about Joseph than about some of his brothers, but some of William's diary entries create the impression that Joseph lacked the drive and shrewdness possessed in such abundance by his brothers. He managed the day-to-day operations of the brewery for a few years after his father's death, but turned them over to William when he came of age and concentrated his efforts on managing the grist mill and farm.

Unlike his Anglican brothers, Joseph was Methodist, likely as a result of the influence of his first wife, Sarah Glassco. One of her diaries has survived. As the following exerpt illustrates, it dwelt almost totally on her spiritual development rather than her day-to-day family life.

> January 6th – Praise the Lord oh my soul, for the renewed tokens of His divine favours again. The refreshing gales of His holy spirit wafted my earthly minded soul towards the heavenly caravan the haven of eternal repose where all is pure unselfish joy.[147]

The youngest Helliwell brother, Charles Lord, was only involved with the Don Brewery for a few years, coming of age and becoming a partner in 1838. He married three times and had approximately 21 children. He worked for awhile at the Eastwood stationery store on King Street in Toronto and later leased and managed the Crooks paper mill in Flamboro.

William Helliwell, whose house still stands at the Todmorden Mills Museum, came of age and became a partner in the brewery in 1832. His first assignment was to travel to England and learn as much about the brewing business there as he could. His diary from that trip has survived and is full of marvellous accounts not only of the various breweries and other industries he visited, but also of some of the tourist and other attractions he saw. His writing style is always fascinating – his spelling even more so! Some of his descriptions are almost too explicit, as the

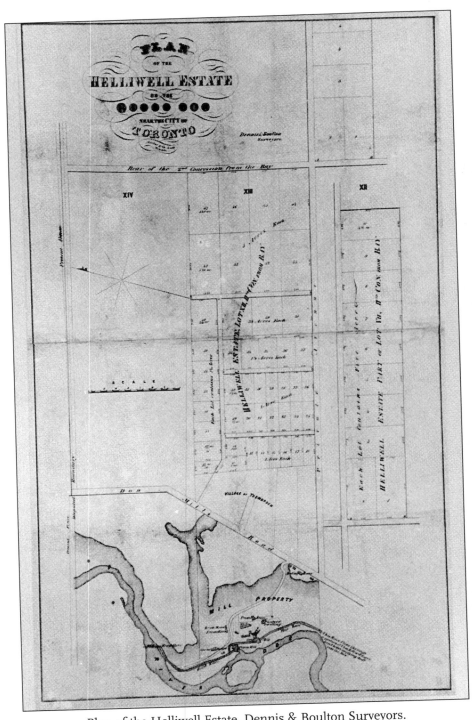

Plan of the Helliwell Estate, Dennis & Boulton Surveyors.
Todmorden Mills Heritage Museum and Arts Centre

following excerpt demonstrates:

> *June 21, 1832 - Fryday being market day at Smithfield I thought I would take a peep at what was going on there. After inquiring twisting and turning through a labrynth of Streets I at last came in sight of a large square of some five or six acres all paved with stone but about half shoe deep with dung & slush as there had been a heavy shoure of rain this Morning. There is ranges or rails to which are fastened Rows of Mooley Bullocks in hundreds & at another side of the Market is pens full of Sheep and lambs and still further on are herds of swine and still further on are hundreds of calves from two weeks to two months old. Such a noise of cows bodeyes Bullocks roaring Sheep bleating pigs squeeling and calves bawling & Buchers and drovers swearing & dogs barking that one could scarseley hear themselves speak ... It came on to rain while I was there which made it aney thing but a comfortable place for it was impossible to stir without being covered with dung and bladder and no attention is paid here wether you hav a clean coate on or not.*[148]

During this trip to England, William took the opportunity to return to his old home in Todmorden where he visited a number of relatives. He discovered, like many who recall places from their childhood, how much smaller and less impressive they seemed to his adult eyes.

William married twice. His first wife was Elizabeth Bright whom he married when he was 23 and Elizabeth 19. The following extracts from the diary provide a charming picture of their rather up-and-down courtship.

> *6 Oct. 1833 – ... called at Mr. Brights and found my Dear Betsey well and as glad to see me back again as I was to see her. I thought I never saw her look so Lovely as then.*
>
> *18 Nov. 1833 – I went to see My dear Girl and stopt till half past nine oclock. Really the time I spend with her is the most pleasant of my life.*
>
> *25 Nov. 1833 – ... in the afternoon I went to Mr. Brights to see Elizabeth and during the evening we had rather a misunderstanding which originated in some questions I asked her which she did not like however we made it up before I left.*

29 Nov. 1833 – At dark I took the horse and road to York to see the dear Girl of My Heart and spent three houres with her verey much to My satisfaction. It is a great effort to leave her so fascinating is her companey to me.

19 Dec. 1833 – I spent three or four houres with the dear Girl Greateley to my satisfaction. Indeed she never was so familiar and free with me as this night.

25 Dec. 1833 – I stopt till all the companey left and fiew minuets after and would hav stopt longer but when Mrs. and Miss Nansey went out of the room I requested Miss Elizabeth to shut the door and she would not so I took my coate and hat and came out of the room in rather a bad humour indeed I was quite vexed.

6 Jan. 1834 – Miss Elizabeth and me had a fiew words about some libertys I took with her.

23 Janu. 1834 – I called at the old place at the cornor of Princess Street where I spent two or three houres with the dear little Girl of My Choise.

28 Jan. 1834 – In the afternoon I went to York and called at Mr. Brights and stopt there till ten oclock. It was my intention to have asked him for his Daughter but as

Mill cottages in the Don Valley. These now vanished structures are very similar to architectural drawings prepared for Joseph Helliwell.

Todmorden Mills Heritage Museum and Arts Centre

A Mill Should be Build Thereon: A History of Todmorden Mills 72

I had not an opportunity I wrote a note to him on the subject and left it for Miss Nancy to deliver.

29 Jan. 1834 – Mr. Bright came up this afternoon to look at some cattle and told me that he was verey willing for me to has his daughter. So if all is well we will be married on the sixth of February.

2 Feb. 1834 – I suppose this is the last Sunday that I am to be a Single Man and God grant that I may be a good husband and she a wife.

6 Feb. 1834 – Wedding Day got married by the Rev'd Doctor Phillips at five PM took supper and <u>brot my Bride Home</u> at twelve and in the one horse waggon.

William got most, but not all, of his wish. The marriage was fruitful and happy, but short. Elizabeth died in 1843, leaving behind four daughters and two sons. William's diary entries about the birth of these children seem the very models of Yorkshire matter-of-factness. The following entry is that he wrote concerning the birth of his first son.

14 March 1839 – This morning I road the colt out to the pine bush. In the afternoon I was chopping wood until

One of the mill dams on the Don, late nineteenth century.
Todmorden Mills Heritage Museum and Arts Centre

*five oclock When Mrs. Helliwell getting so ill (she had
been complaining all day) I walked to city (as we had no
horse at home but the colt) for Dr. King who arrived
before I did. I had not been maney minutes at home
before I was informed that Mrs. Helliwell was delivered
of a verey fine Boy.
I then took the Horses and waggon and went for Miss
Bright [his sister-in-law] the news pleased me.*

The years from 1842 to 1844 were filled with losses for William. His
mother died in 1842; his wife in 1843; and then in 1844 his second son,
William, was killed in an accident at the brewery, possibly drowning in the
mill-race. Left alone to care for six children under the age of nine, William
needed a housekeeper. This role was filled by his sister-in-law, Jane, whom
he married in 1844. They had eleven more children.

The entire Helliwell family seems to have been an extremely healthy
lot, especially considering the mortality rates normal at the time. Only one
of William's seventeen children died in infancy – Nellie, who was born
when Jane was 44 years old. William lived to be 86, dying in 1897. His second wife, Jane, lived to age 83.

William was a man of considerable ability and curiosity, keenly interested in the history of the area and an interested observer of the events of
his day. He made one of the first measured surveys of the Don, helped
construct the second road into the area and became path master of the
Don Mills Road – all before the age of 30. His account of his role as overseer for the construction of the area's second road provides an interesting
insight into how the community worked together to obtain their needs.

*In the year 1841 the inhabitants of the Don became dis-
satisfied with the toll gate placed at the Don Bridge, in
consequence of as much toll being charged for teams coming down the Don Mills Road and travelling less than
half a mile on the Kingston Road as those travelling five
miles on said road and began to look about for some
other way of reaching the City and having procured permission from John Scadding to open up and construct a
road leading off from the Don Mills Road at a point
where the above mentioned road crosses the first concession of York where a natural gully exists leading down*

the high bank to the Don River. In December 1841 the
inhabitants turned out with men and teams and com-
menced the construction of a road down to the Don
under my supervision and continued at it from day to
day as voluntary labour until a passable road was made
down to the Don and up the hill on the west side bound-
ing what is now the Necropolis. During the winter tim-
ber was procured and preparations for erecting a bridge
over the river and in the spring the road was opened for
travel passing down Sumach Street to King Street.[149]

The condition of the area's roads was a constantly recurring complaint in William's diaries. During some periods of the year it was almost
impossible to get the heavy wagons of beer into town. William records one
time when they had to remove all but two kegs from the wagon before the
horses could reach the top of the hill above the mills. Scadding described
a method that they used at such times in his account of one of the Don
Valley's more unique residents of that period.

Not far from the spot where, at present [1873], the Don-
street bridge crosses the river ... lived for a long time a her-
mit-squatter, named Joseph Tyler, an old New Jerseyman
of picturesque aspect ... His abode on the Don was an
excavation in the side of the steep hill, a little way above
the level of the river-bank ... He built for himself a mag-
nificent canoe, locally famous. It consisted of two large
pine logs, each about forty feet long, well shaped and deft-
ly hollowed out, fastened together by across dove-tail
pieces let in at regular distances along the interior of its
bottom ... In this craft he used to pole himself down the
windings of the stream ... He would also on occasion
undertake the office of ferryman ... At the season of the
year when the roads through the woods were impractica-
ble, Tyler's famous canoe was employed by the Messrs.
Helliwell for conveying into town ... the beer manufac-
tured at their Breweries on the Don.[150]

Even by the 1850s the Don Mills (or Todmorden Mills as they were now
beginning to be called) were still quite rural, despite their new roads and

industries as the following account from William's biography illustrates.

> ... the household at the Don was disturbed in the middle of the night by the howling and crying of the dog, a large and at times rather fierce mastiff but now was completely frightened and crouched against the door. On opening the door was seen four dead and dying sheep that had just been killed by the wolves. In the morning we found the sheep everywhere and completely stunned. The same night our neighbour Samuel Sinclair lost fourteen sheep, no doubt killed by the same pack of wolves.[151]

The village of Todmorden was still, of course, quite small. A map of the area prepared in 1855 shows only four houses, the mills, some barns and assorted out-buildings in the valley and only two structures at the top of the hill above.[152] One of these upper buildings was labelled as being a "meeting-house." The Helliwell family home had been burnt before the date of this map, and there must have been some workmen's housing not shown, but even so, it was a very small community.

Before it burned, the Helliwell family home was a multi-storey stone structure attached to the 'cooler' portion of the Brewery. Some of the workmen slept in its top floor. This was the Helliwell's second home and had replaced their original frame house. Joseph and his family lived there with his elderly mother, although she often stayed in Toronto with her other children. William and Elizabeth also lived there during the first few years of their marriage until their own home was finished a little further up the hill (around 1838). Their new home consisted of a main two-storey portion constructed of unfired mud bricks, stuccoed on the exterior, with a one-storey frame wing. Marks found by the museum's restoration architect on the sections of mud-brick (adobe) where the wing adjoined it, indicated to him that this original wooden portion had burned and been replaced around 1850.[153]

Only little scraps of information exist about the lives of the Helliwell workmen. William refers to a few of them by name in his diary, but most are referred to only by their occupation, e.g. the maltster. The museum archives contain an architectural drawing of workmen's houses that was prepared for Joseph Helliwell, but it is not known where or even whether these houses were ever built.[154]

A somewhat unusual source does provide us with information concerning their pay. In 1847 the gaoler in York asked for a raise. This was refused by the Magistrates who stated in justification that "Messrs. Eastwood and Hallowell pay to their brewers, etc., only eight shilling and nine pence a week, and that [to] the houses in which they are boarded [they] pay house rent, firewood, etc. and make a decent living."[155]

From a later biography about his son who became a prosperous wagon-maker, we learned the name of one of their workmen, John Smith, and something of the trials faced by early immigrants.

> In 1832, his [William Smith's] parents, John and Mary (Mason) Smith, came to Canada with their children. On the voyage out, on the ship Alexander, the smallpox and cholera broke out among the passengers, and Mr. Smith lost a sister, two years old by the former, and while waiting at Prescott for a boat to take them to York, the mother died of cholera, in a shop there, leaving the father to look after three young children. Mr. Smith's uncle and grandfather died at Montreal of the cholera. After reaching

The Helliwell House - 1838.

Helliwell House, 1838. John Ross Robertson, *Landmarks of Toronto*, Vol. 1, 1894,
Metropolitan Toronto Reference Library

York his father rented two rooms on Yonge Street, and obtained work as a mason's clerk. He afterwards became a labourer in Helliwell's Brewery and died in 1849.[156]

Relations between the Helliwells and their brothers-in-law, Eastwood

William and Joseph Helliwell.
Todmorden Mills Heritage Museum and Arts Centre

and Skinner, became increasingly strained during the 1830s as their disputes over the site's mill-races and the boundary lines of their properties grew, until they could only be resolved in court. When Thomas Helliwell Sr. bought land from Eastwood and Skinner in 1822 for the purpose of constructing a raceway off their mill-race, the issue of who had primary rights to the water were not made clear. This became a point of contention whenever the water level dropped. Their agreement had also failed to specify who was responsible for clearing ice and debris from the mill-race, and hence who should pay compensation when it caused damage. The courts eventually supported the position of Eastwood and Skinner, upholding their claim that they had primary control over the mill-race as its first owners and users.

There were also disagreements over the correct boundary lines of the property that Thomas Sr. had purchased. The dispute escalated when Eastwood and some of his men tore down the fence that the Helliwells had erected along the boundary and they took him to court. He was fined £20 for tearing down the fence, but the boundary remained unsettled for years. The 1855 map mentioned earlier still labelled the boundary line as "disputed."

These bad feelings within the family were further exacerbated by their opposing political positions. Eastwood and Skinner favoured the Reformers although they would not support rebellion, while all of the Helliwell brothers worked to elect 'Family Compact' members such as Sheriff Jarvis and were staunch supporters of the establishment.

Although most of the conflicts during the 1837 Rebellion happened along Yonge Street, the Don Valley did see some fighting. On the final day of the Rebellion, a group of rebel supporters under the command of Peter Matthews, were sent to burn the Don Bridge and thus cut off the possibility of government reinforcements reaching the city from the east. They succeeded in setting fire to the bridge, but the fire was soon extinguished by the city's volunteer fire department. At the same time, government forces succeeded in routing the rebels in a short battle near Montgomery's Tavern and Mackenzie and his supporters were forced to flee. Peter Matthews was caught, found guilty of treason and hanged the following year.

According to the autobiography of Captain Philip DeGrassi, a local resident of the time, Matthews and his men had not gone directly to the bridge. He wrote that "On the evening of the day that the Rebellion broke out, as a well-wisher of law and order, I went to Toronto to offer my services to the Government, accompanied by my two daughters, and narrowly escaped being taken by Mathew's [sic] troop, which was going to Mr. Helliwell's place, where, although unwelcome guests they regaled themselves."[157]

This same source gives the Don Valley its claim to a loyalist heroine during the Rebellion.[158] DeGrassi's autobiography made much of the incident since it formed part of his complaint that neither he nor his daughter had ever received any reward for their services to the government. Mackenzie confirmed the basics of the story in his account of the Rebellion, but only briefly and in passing. Whatever the actual significance of the incident, it is still a good dramatic story.

> I [DeGrassi] said that I would endeavour to ascertain the number of the rebels on Yonge Street. One of my daughters about 13 years of age, accordingly, who was a capital rider rode out under pretence of wishing to know the price of a sleigh, went to a wheelwright's shop close to Montgomery's Tavern, and being suspected, was taken prisoner by some of the rebels who ordered her to dismount. To this she demurred and during her altercation with her captors MacKenzie came with the news that the Western Mail was taken. Amidst the general excitement my little girl had the presence of mind to urger [sic] her horse and ride off at full speed amidst discharges of musketry. A ball went through her saddle and another through her riding habit.
>
> Arrived in Toronto she was taken before Sir F.B. Head, the Governor to whom she gave valuable information as to the numbers and condition of the rebels – thus the loyalists were encouraged, measures were taken to meet MacKenzie's attack and so my poor child was the means of saving Toronto where he had many partisans.[159]

Captain Philip DeGrassi came from a distinguished background. His father had been a captain in the service of the Cisalpine Republic, one of the puppet states set up by Napoleon in his conquered territories. His mother was the daughter of an Austrian baron and his grandmother had been a lady-in-waiting to the Queen of Naples. Philip DeGrassi began his military career in Napoleon's army, but was captured by the British and changed sides. After the Napoleonic Wars he was retired on half-pay and moved to Chichester, England. He was fluent in at least five languages, so taught languages to the nobility of the community to supplement his income.

Advised by his friends to better his fortunes by taking up land in Canada he chartered a boat for his family and emigrated with his wife (who was going blind) and seven of their children. He brought with him about £3,500 and "ploughs quite unfit for the country, a portable flour mill, 36 cow bells, 600 sheep bells, carpenter's and blacksmith's tools, and as the auctioneers say "many other items too numerous to mention." He later admitted that "Many of them [were] so incongruous as to be almost useless" and that "I was full of the idea of being a farmer, although I was profoundly ignorant of the art of agriculture."[160]

When he arrived in York he was offered several different town lots, but instead chose to accept a larger lot on the Don River north of Todmorden. He fulfilled the obligations to clear land and build a home and was confirmed in his grant. He bought the neighbouring property and built a saw-mill, but failed to prosper. He had no experience in any of these endeavours and therefore had to hire workmen. His vision of supporting himself as a landlord in the manner of the country gentry of England and Italy was not possible in Canada. Some of his lifestock were stolen and he was cheated by his miller. To cap this litany of troubles, his house and its contents were destroyed by fire and his family were forced to live in the barn where "one of [his] children was literally born in a stable and laid in a manger."[161] The DeGrassi sons remained in the Toronto area. Both of his daughters married Americans and moved to the United States. Philip DeGrassi died, bitter and impoverished, in 1860.

William Helliwell's part in the Rebellion was considerably less dramatic. His diary records that he returned by boat from Niagara about 1:00 on December 5, 1837 and learned at that time that:

> ... the Radicals had made an attempt to take the city and had taken several of the citisens prisoners and had shot Colonel Moody ... and was expected in every moment to sack the city. On going up to the market I found every shop shut and the inhabitants all in arms with field pieces placed in the streets. Thomas [his brother] was in arms at the Bank and had been there since two oclock of the preceding night. On coming down to Mr. Brights [his father-in-law] heard that the Radicals had fired Dr. Horns House and on looking up I distinctly saw the

> colum of smoke raising. Truly the city was very much
> alarmed ... There is not a doubt on my mind that if the
> Rebels had pushed an assault the Terror and confusion
> of Monday night they would have got possession of the
> arms and amunition and with them the city. And I can
> hardly contemplate the consequences that would have
> ensued from such a misfortune. The city would hav been
> a prey to every species of plunder and repine.[162]

He wrote that he returned home to the brewery and that night, "30-40 men armed with Rifles" bound for Yonge Street used a log to cross over the Don near the brewery, presumably to join Mackenzie's forces. His entry for December 6 notes that he went back into town with the workmen from the brewery but returned home that evening since nothing had occurred. On December 7 he went back again and "was sworn in as a special constable and went up to the Parlement Building to get arms ... [and] amunition when Mr. Small pressed me to stop and give out powder and shot."[163] He continued with this task until three o'clock when he was sent to a plumber's shop to make musket balls. He recorded with some pride that he made "two pails full" of them. At nightfall he went home, encountering some of the returning men who were accompanying the carts of wounded.[164] On December 10, he went "to see the battle grounds and the ruins of Montgomery's House which was burnt on that ocation. I took notice of one House that belonged to a leading Reble that was pearsed with several shot so much so that I should hav been verey sorry to hav been inside."[165]

William became a captain in the North York Militia in early 1838 and his diary entries for that year are full of accounts of the drills and other military exercises he engaged in with that unit. By 1839, however, the sense of alarm had diminished and there were fewer drills and shorter exercises. Both usually ended at a local tavern where (in William's words) all present parted "in good feeling and friendly union at a very early houre."[166]

William remained embittered against the rebels for a long time. His diary entries show that he was personally offended (and probably very worried) that the rebels had forged his name to an address they sent to the Governor. His uncomplimentary descriptions of the rebels and their American allies in his diary as "the Pirates from across the Lake" and "the Mob of America" give some indication of the depth of his feelings. He declaimed "Let them come if they dare. They will not all return." He also recorded, with considerable relish, the details of an attempted suicide by the rebel General Sutherland and recorded his attendance at the trials of Samuel Lount, Peter Matthews and John Montgomery. His entry for April 12, 1838 is full of real regret that he "went to the city expecting to see Lount and Mathews hung but was too late they being cut down about half an houre before I got there."

The Helliwell businesses continued to prosper until their calamitous losses from a fire ended the partnership in 1847. Sarah Glassco Helliwell, Joseph's wife, departed from the usual purely religious content of her diary to describe the fire.

> The alarm of fire was Givin Just as we were retiring to bed part of my family was already in bed as also our men who slept in the top story of the house. At the thrilling cry of Fire my dear Joseph and my self pushed to the door when to our utter astonishment and dismay the Coolers from which there was a communication to our house was all on Fire. I felt conscious our dwelling house could not be saved and so it proved for I believe in less than two hours the brewery, dwelling house and Grist Mill were all

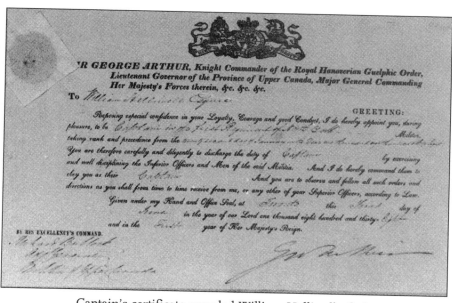

Captain's certificate awarded William Helliwell, 1838.
Todmorden Mills Heritage Museum and Arts Centre

A Mill Should be Build Thereon: A History of Todmorden Mills 83 *consumed the loss of which can scarcely be estimated as it is not only the loss of property but has caused a complete stagnation in the business. Thus the infinitely wise God my heavenly Father hath been pleased to permit a heavy gloom to shade the sun of prosperity.*[167]

The *Toronto Globe* report on January 13, 1847 said the cause of the fire was the "putting of ashes in a wooden vessel" next to the brewery. A later source declared that the "loss was estimated at about $80,000 of which $5,000 only was covered by insurance. Many of the workmen employed by Mr. Helliwell had all their clothes burnt, and all had a narrow escape from being burnt to death, as the stairs in the house where they slept were consumed before they woke."[168] According to John Ross Robertson, the brewery was never rebuilt and the burnt stone walls were covered with a roof and used as a storehouse.

The partnership between the Helliwell brothers was dissolved and the remaining properties and money divided between them. Thomas and Joseph retained the Don property while William took the land they had acquired in the Highland Creek area of Scarborough. Charles Lord took his share in cash.

View looking north-east across Helliwell's wharf to the rear of City Hall, 1849.
Pencil drawing by F.H. Granger.
Metropolitan Toronto Reference Library T-10347

In Highland Creek, William became involved in a variety of businesses and speculations. He built a saw-mill and a grist mill there and later built a hotel. He held shares in a lake cargo boat and in the abortive Scarborough Oil Company. He served as a municipal Councillor, a Justice of the Peace and as Commissioner of Fisheries for the area. Although generally successful, he had to endure a number of serious financial blows. His grist mill burned in 1879. His house also burned soon after and he moved his family into the hotel.

According to the family record, Joseph lost most of his money to bad investments in the United States, but held onto most of the valley farm land. He also built and operated a new grist mill near the site of their earlier one. Thomas continued to live and work in York, maintaining his involvement with the financial community. Both Thomas and Joseph slowly sold off their property on the Don to the Taylor family, beginning with the land immediately around the brewery. By 1855, the Taylors had acquired virtually all of the old mills along with their dams, mill-races and mill houses.

The Taylor family bought the property near the present Leaside Bridge and moved to the Don Valley in the early 1830s. They prospered greatly over the next fifty years, eventually owning approximately 2,000

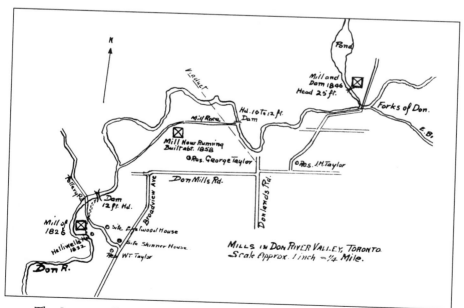

The Lower Mill was the one originally owned by Eastwood & Skinner.

acres, three paper mills and a brick factory in the Don Valley and surrounding East York area. Although several of the Todmorden Mills continued to operate for many decades, their unique history ended when they were absorbed into the larger Taylor industrial complex. Time and progress took their toll on the mill property so that today it is only with knowledge and imagination that one can begin to imagine the bustling village that once stood there.

EPILOGUE

The land originally known as the Don Mills and later Todmorden Mills experienced many changes throughout the 19th and early 20th century. By 1855, after the Taylor brothers purchased the Eastwood paper mill, the mill site entered a new era of industrial use. As one of three Taylor owned paper mills, the mill at Todmorden produced felt paper and became known as the Lower Mill.

On November 12, 1900 the Lower Mill was damaged by fire and its subsequent history remains unclear. Robert Davies bought the Lower and Middle mills in 1907 and acted as owner until his death in 1916. By 1928 the properties held by the Davies estate were sold and it is believed that the existing mill building became an integral part of the Don Valley Brickworks operation. Horses used at the brickworks were stabled on site, and damaged brick was dumped on the surrounding lands.

The precise use of the original site buildings at Todmorden Mills is unclear during the latter part of the 19th century and into the 20th century. We do however know that the Paper Mill served as a riding stable in later years and the historic houses continued to provide private residential housing until 1965.

During the years leading up to the commemoration of Canada's centennial year in 1967, the Township of East York prepared to make plans for a local celebration. The community's excitement sparked by the approaching celebration was further enhanced by the observation that:

> Higher governments will pay $2 to every $1 spent by the Township on some special project of a lasting nature to commemorate Canada's One Hundredth Birthday. They suggest parks, buildings of historic interest with historic furniture, compilation and publication of histories, collection of works of art, and other records of the development of our country and our locality. (*East Yorker*, October 1964, no. 15 p. 1)

By 1964 The East York Centennial Committee was established and the existing Todmorden Mills Park was seen as a most fitting project. It not

only satisfied the criteria for projects, but there were four historic buildings at Todmorden Mills still on their original location! The site consisted
of Parshall Terry's early 19th century home, William Helliwell's adobe
brick house, built in two differing periods, 1821 and 1837-38, the Paper
Mill and remnants of what was at that time considered to be the 1847
brewery building.

An act of the Provincial Legislature in 1965 established the East York
Foundation, which had as part of its mandate, fundraising for the preservation of the historical buildings remaining at the Todmorden Mills Park.

As the project progressed, Peter Stokes was hired as consulting
restoration architect and Mrs. Nell Donaldson was hired to assist in fur-
nishing the historic houses. Interested members of the public were actively involved in collecting and documenting their local history. The junior
section of the East York Horticultural Society, it was stated "is planning to
plant a garden of authentic, old-fashioned flowers and vegetables beside
one of the houses, and it is suggested that the wilderness groves be preserved and wild flowers be re-introduced along the pathways." (*East Yorker*,
October 1964, no. 15 p.1.)

On May 22, 1967 the Todmorden Mills Historic Site opened to the
public as the Borough of East York's community museum. At the prompt-
ing of Mayor True Davidson, a group of dedicated volunteers known as the
Todmorden Mills Guild were recruited to provide assistance with cos-
tumed interpretation, public events at the site and the day to day operation
of the museum.

Reflecting upon the early beginnings of the site's evolution into East
York's community museum, it is interesting to note the correlation with
present-day initiatives at the museum. In 1995, the Todmorden Mills
Heritage Museum and Arts Centre is actively involved in diverse projects
to serve a public interested in industrial, social, natural and local history.
The Todmorden Mills Wildflower Committee is a non-profit group devot-
ed to naturalizing the area surrounding Todmorden Mills through the
introduction of native species of trees, plants and shrubs. The museum's
Heritage Garden Committee has recently received support from Canada
Trust's Friends of the Environment Foundation to develop phase one of a
landscape plan commissioned by the Leaside Garden Society. Any reflec-
tion upon past history and the thoughts and aspirations of those associated with Todmorden Mills today only reinforces the fact that history does
indeed repeat itself.

Initially the museum was staffed by volunteer curators who provided their services in exchange for a small honorarium. By 1974 Eleanor Darke was hired as a full time curator and she began to piece together the substantial and detailed history of the site spanning 200 years. Her work has assisted present-day staff in documenting the site's complex historical development. It has also raised many more questions and has exposed areas which have been the subject of further research. We now know that the house attributed to Parshall Terry dates somewhat later than originally thought, and that the brewery building may well be an early 20th century structure, although vestiges of the second brewery on site dating to 1835 may remain in archaeological terms.

There is no question that the many individuals who contributed to the development of Todmorden Mills historically have left us with a rich and varied history. In 1993 I received funding from the Canadian Museum Association and The British Council to travel to Todmorden, Yorkshire in search of some answers pertaining to the Helliwell family and their emigration to Upper Canada. The information collected during my visit to Todmorden, Yorkshire provides a small but fascinating piece of the complex and often enigmatic story of the history of East York's Todmorden Mills.

The importance of providing continuous support in the preservation and interpretation of Todmorden Mills is further enhanced with recent developments of the Don Valley Brickworks. Funds have been secured to offer protection, enhancement and rehabilitation to the originally established 1890 Taylor-owned Don Valley Pressed Brick Company. Long term plans may include a physical link joining the Don Valley Brickworks site to Todmorden Mills.

Today, the Todmorden Mills Heritage Museum and Arts Centre remains a site that has undergone relatively little physical disturbance and change since its development as a public museum. In its predominantly pristine condition the site offers tremendous potential in both archaeological and interpretive terms. The buildings have been designated under the Ontario Heritage Act, and the area of Todmorden Mills has been identified by the Metropolitan Toronto and Region Conservation Authority as a site of outstanding heritage value within the Don River Watershed. (Don Valley Watershed Heritage Study, 1933, p. 19)

The future will hold many new and exciting prospects as the industrial story that the Don River Valley has to tell unfolds with recent developments and linkages between the Don Valley Brickworks and Todmorden Mills. The resurgence of public interest in industrial history

will undoubtedly help to ensure that this region of the Don Valley is preserved and interpreted by present day custodians for the benefit of a future
generation.

Susan Hughes,
Curator/Administrator
Todmorden Mills

Bibliographic references:
1. *East Yorker*, Borough of East York, Number 15, October 1964.
2. The Don Watershed Heritage Study, The Metropolitan Toronto and
    Region Conservation Authority, 1993.

The Parshall Terry House is one of the oldest houses on its original site in Metropolitan Toronto.

Boris Novikoff, East York

The William Helliwell House was built in two stages and finished in 1838.
It has been expertly restored and contains period furnishings.

Boris Novikoff, East York

Built in 1821, the Helliwell Brewery once supplied York inhabitants
with the finest in ales. Today, the building houses museum office staff
and displays.

Boris Novikoff, East York

ENDNOTES

Chapter 1 – The Don Valley

1. Geological information obtained from "Sections in the Don Valley," unpublished article, author unknown, Todmorden Mills Museums archives.

2. James Helliwell, editor, grandson of William Helliwell. "Diary of William Helliwell," unpublished m.s.,Todmorden Mills Museum Archives, compiled before 1933

3. John Ross Robertson, ed., *The Diary of Mrs. John Graves Simcoe*, (Toronto, 1911), p. 335.

4. John Ross Robertson, ed. *Landmarks of Toronto, Volume VI* (Toronto, 1894), p. 212

5. Henry Scadding, *Toronto of Old*, (Toronto, 1873), p. 220

6. Robertson, *Landmarks of Toronto, Vol. VI*, p. 212

7. J. E. Middleton, *The Municipality of Toronto, Volume II*, (Toronto, 1923), p. 751

8. Robertson, ed., *The Diary of Mrs. John Graves Simcoe*, p. 298

9. Middleton, p. 751

10. Hugh McCrimmon, "Salmon in the Don River," Todmorden Mills Museum archives.

11. Robertson, *Landmarks of Toronto, Volume VI*, p. 212

12. Scadding, p. 220

13. Robertson, ed. *The Diary of Mrs. John Graves Simcoe*, p. 335

14. Ibid, p. 335

Chapter 2 – The Valley's First Inhabitants

15. Percy Robinson, *Toronto During the French Regime*, (Toronto, 1965), p. 97

16. Robinson, p. 98

17. Robinson, p. 98

18. James Helliwell, ed., "Diary of William Helliwell"

19. Charles Sauriol, *Remembering the Don*, (Toronto, 1981), p. 141

20. Sauriol, p. 141

21. Scadding, p. 233

22. Robinson, p. 165

23. "Report by Alexander Aitken to the Hon. John Collins, Deputy Surveyor General, from Kingston, Sept. 15, 1788," Public Archives of Ontario

24. E.A. Cruikshank, ed., *Correspondence of Lieut. Governor John Graves Simcoe*,

Volume II, (Toronto, 1923), p. 61

25. E.C. Guillet, Toronto, (Toronto, 1934), p. 4.
26. Cruikshank, *Correspondence of Lieut. Governor John Graves Simcoe, Vol. II,* p. 304
27. "Russell Papers – Chief Justice Elmsley to Russell, Feb. 2, 1797," Toronto Public Library, p. 69.
28. James Strachan, *A Visit to the Province of Upper Canada in 1819,* reprint, (Toronto, 1968), p. 145.
29. "Russell Papers – Letter from Peter Russell to John King, February 1, 1797."
30. Edwin C. Guillet, *Early Life in Upper Canada,* (Toronto, 1963) p. 385
31. Edith Firth, *The Town of York 1793-1815,* (Toronto, 1962), p. 88.
32. Joseph Andre, *Infant Toronto,* (Toronto, 1971), p. 74-75.
33. Letter from William Jarvis to Alexander Macdonell, June 30, 1797, p. 506
34. Andre, p. 74-5
35. V.B. Blake, *Don Valley Conservation Report,* (Toronto, 1950), p. 84.
36. Firth, p. 88
37. John Ross Robertson, *History of Freemasonry,* Volume II, (Toronto, 1899), p. 316.

Chapter 3 – The Don Mills

38. "Wolford Simcoe Papers," Volume II, Toronto Public Library, p. 231
39. "John McGill Papers – McGill to Hunter, 16 Oct. 1799," Public Archives of Canada, p. 51.
40. Ernest Green, "Some Graves in Lundy's Lane," *Niagara Historical Society Journal,* No. 22 (Welland, 1911), p. 55.
41. James W. Burbank, *Cushetunk: 1754-1784,* (Callicoon, New York, 1952), p. 10-11.
42. Burbank, p. 10-11
43. Burbank, p. 22-23
44. James Moody, *Lt. James Moody's Narrative of his Exertions and Sufferings,* reprint. (New York, 1968).
45. "American Loyalist Transcripts," Vol. 16, Audit Office, Public Record Office, London, England, p. 559
46. "Loyalist Transcripts," Vol. 16, p. 559
47. "Loyalist Transcripts," Vol. 28, p. 595
48. "Upper Canada Land Petitions, 'S' Bundle Miscellaneous, 23 Dec. 1811," Public Archives of Canada.
49. Frank N. Walker, *Sketches of Old Toronto,* (Toronto, 1965), p. 12
50. "Wolford Simcoe Papers," Volume III, p. 365
51. Blake, p. 70
52. "Letters to the Surveyor General, D.W. Smith, from Alexander Aitken, Deputy Surveyor," 4 June 1795, Public Archives of Ontario, p. 902.

53. "Wolford Simcoe Papers – Meeting in the Council Chamber York 20th July 1796," Toronto Public Library, p. 198.

54. "Wolford Simcoe Papers – Voucher Entries by John McGill, Commissary of Stores, 14th July 1796," Toronto Public Library, p. 363.

55. E. A. Cruikshank, editor, "Simcoe Papers," *Ontario History*, Volume 26, p. 331.

56. William Lea, "The Valley of the Don," *The Evening Telegram*, Feb. 4, 1881

57. James Helliwell, ed., "Diary of William Helliwell."

58. Pauline Reaburn, "Power from the old mill streams," *Canadian Geographical Journal, Volume 90*, March 1975, p. 23.

59. "Travels of Francois Alexandre Frederic la Rochefoucault-Liancourt to Upper Canada in 1795," *Ontario Archives 13th Report*, (Toronto, 1916), p. 24.

60. Reaburn, p. 25

61. "Abner Miles Account Books," May 11, 1796, Toronto Public Library

62. "Eli Playter Diary," April 1804, Public Archives of Ontario

63. "Home District Affidavits and Dispositions Miscellaneous," RG 22, Series 7, Volume 35, Public Archives of Canada.

64. IBID

65. E.C. Kyte, *Old Toronto* (Toronto, 1954), p. 41.

66. William Chewett, "Plan of ... The Township of York", Metropolitan Toronto Reference Library

67. *Proceedings of the New Jersey Historical Society*, p. 474.

68. Robertson, ed., *Diary of Mrs. John Graves Simcoe*, p. 335

69. "John McGill Papers – George Playter to McGill, March 15, 1797," Public Archives of Canada.

70. "Minutes of Town Meetings, 1797," Metropolitan Toronto Public Library

71. Ibid

72. William Lea, "The Valley of the Don," *The Evening Telegram*, Feb. 4, 1881.

73. Joseph Willocks, ed., *Upper Canada Guardian*, June 22, 1810.

74. K.R. Macpherson, *The King's Mill: A Resume* (Toronto, 1963), p. 18.

75. "Hunter Papers – Aeneas Shaw to P. Hunter, 24 May 1805," Public Archives of Canada.

76. "Upper Canada Land Petitions, 'S' Bundle Miscellaneous, 21 Sept. 1818," Public Archives of Canada.

Chapter 4 – Parshall Terry

77. Kyte, p. 217

78. Norah Hall Lund, *Parshall Terry Family History* (Paragonah, Utah, 1956).

79. "Item 132, List of Rioters at Fort Wyoming, 21 Jan. 1771," Susquehannah Papers (Wilkes-Barre, 1933), p. 154.

80. "Colonial Office Records, Volume 13, MG 11, 'O' Series – A List of Persons Employed as Rangers in the Indian Department, June 15, 1777," p. 331.

81. "Susquehannah Papers," p. 154

82. "Haldimand MSS, B 126 – Roll of Loyalists at and in the Dependencies of Cataraqui and the Provisions They Draw," Public Archives of Canada, p. 91-5.

83. E. A. Cruikshank, ed., "A List of Persons who have Subscribed their Names in Order to Settle and Cultivate Crown Land," *Ten Years of the Colony of Niagara, 1780-90*, Appendix C (Welland, 1908), p. 45.

84. E. A. Cruikshank, "Settlement of the Township of Fort Erie," *Welland Historical Society*, Volume 5 (Welland, 1938), p. 25.

85. C.C. James, "First Legislators of Upper Canada," *Royal Society of Canada Proceedings and Transactions*, 1902, p. 108-118.

86. "Wolford Simcoe Papers – Ralph Clench to Major Littlehales, July 8, 1796," Volume 6, Toronto Public Library, p. 292-3

87. "Upper Canada Land Petitions, 'T' Bundle, Item 5a, 25 Aug. 1801"

88. Robertson, ed., Landmarks of Toronto, Volume VI, p. 250

89. *Upper Canada Gazette*, June 14, 1800.

90. J.E. Middleton, "Diary of Joseph Willcocks," *History of Ontario* (Toronto, 1927), p. 1250.

91. Scadding, p. 223

92. "Eli Playter Diary," Public Archives of Ontario.

93. "A. Wood Letterbooks – Wood to Cartwright," Toronto Public Library

94. Ibid

95. *York Gazette*, July 23, 1808. Quoted in John Ross Robertson, *Landmarks of Toronto*, Volume VI, p. 357

96. Robertson, Landmarks of Toronto, Vol. VI, p. 357

97. "Lincoln County Wills, 1813-1833 – Will of Timothy Skinner, dated 2nd May 1807, filed 25th September 1815," GS Ont. 1-649, Surrogate Court #579158, District of Niagara, Upper Canada, Will #2198, Public Archives of Ontario

98. Mrs. Ron Glassford, "The Cornell Family," *The Loyalist Gazette*, April 1963, p. 6.

99. Charles Fothergill, *A Few Notes of a Journey from Montreal to Upper Canada* (York, 1817), p. 88

100. Glassford, p. 6

101. Glassford, p. 6

Chapter 5 – A Skinner Returns

102. "James Maclem Papers, 1788-1814 – List of Lincoln Militia, Military Census," Metropolitan Toronto Public Library.

103. "Minutes of Town Meetings – List of the Inhabitants of the Township of York, 1797-1822," Metropolitan Toronto Public Library.

104. "Home District Affidavits and Dispositions Miscellaneous," RG 22, Series 7, Volume 35, Public Archives of Canada.

105. Robertson, ed., *Landmarks of Toronto*, Vol. VI, p. 113

106. "Eli Playter Diary," July 31-Aug. 1, 1813, Public Archives of Ontario

107. "Playter Diary," July 31-Aug. 1, 1813

108. Robertson, ed., *Landmarks of Toronto*, Volume VI, p. 113

109. Rev. John Carroll, *My Boy Life* (Toronto, 1882), p. 264

110. *Don Mills United Church Leaflet for the 154th Anniversary*, Sunday, Nov. 25, 1973, Todmorden Mills Museum Archives.

111. "Lincoln County Wills, 1813-1833 – Will of Timothy Skinner, dated 2nd May 1807, filed 25th September 1815," GS Ont. 1-649, Surrogate Court #579158, District of Niagara, Upper Canada, Will #2198, Public Archives of Ontario

Chapter 6 – Paper-Making On The Don

112. Frederick H. Armstrong, *Toronto, The Place of Meeting*, (Toronto, 1983), p. 63 and p. 97

113. "Will of Timothy Skinner"

114. *York Observor*, August 1822

115. "Instrument No. 6290," Land Titles Office.

116. *Baines Yorkshire Directory*, (London, 1822)

117. Robertson, ed., Landmarks of Toronto, Vol. VI, p. 428

118. "Letter from Alice Eastwood," June 6, 1944, Todmorden Mills Museum Archives.

119. Robertson, *History of Freemasonry*, Vol. II, p. 284

120. George Carruthers, *Paper in the Making* (Toronto, 1946), p. 278-9

121. *The Colonial Advocate*, April 17, 1826

122. *The Evening Telegram*, Feb. 4, 1881.

123. Charles Sauriol, "The Valley of the Don," Todmorden Mills Museum Archives

124. "Appendix, Receiver General's Securities &C., "Report of the Select Committee appointed to consider and report upon the Petition of John Eastwood and Collin Skinner, Paper Makers, at the Don Mills, near York," Journal of the House of Assembly of Upper Canada, 1828-29.

125. Carruthers, p. 299-300

126. Carruthers, p. 299-300

127. Carruthers, p. 299

128. Carruthers, p. 300

129. Carruthers, p. 300

130. "Letter from Miss Edna Ash, Jan. 25, 1983," Todmorden Mills Museum Archives

131. *Upper Canada Gazette*, 1832

132. Armstrong, Toronto, p. 63

133. Kyte, p. 214-5

134. Carruthers, p. 303

135. Carruthers, p. 303

136. Ruby Kinkead, "East York: Century of Education - 1856-1963," Todmorden Mills Museum Archives.
137. Scadding, p. 224
138. W.H. Smith, *Canada Past, Present and Future* (Toronto, 1851), p. 19

Chapter 7 – "A Vigorous and Substantial Family"

139. Scadding, p. 16
140. Glyn Williams, "Calder Valley mill families," *The Illustrated London News*, November 1975
141. James Helliwell, ed., "The Diary of William Helliwell."
142. Albert F. Helliwell, *Helliwell Family Record* (Portland, 1949), p. 28-29
143. James Helliwell, ed., "The Diary of William Helliwell."
144. Robertson, ed. *Landmarks of Toronto*, Vol. VI, p. 429
145. "Old York Wills, 1821-1826 – Will of Thomas Helliwell, July 5, 1823," Public Archives of Canada
146. "Diary of William Helliwell," Feb. 22, 1833, Todmorden Mills Museum Archives
147. "Diary of Sarah Glassco Helliwell, " Todmorden Mills Museum Archives
148. "Helliwell Diary, 1832-33," Todmorden Mills Museum Archives
149. James Helliwell, ed., "Diary of William Helliwell"
150. Scadding, p. 298-299
151. James Helliwell, ed., "Diary of William Helliwell"
152. Dennis & Boulton, Surveyors, "Plan of the Helliwell Estate on the River Don near the City of Toronto," Maclear & Co. Lith., Toronto, 22 Nov. 1855
153. "Notes by Peter Stokes," Restoration file, Todmorden Mills Museum Archives
154. Joseph Sheard, "Designs for a Double Cottage for Joseph Helliwell, Esq., Don Mills, Toronto", July 1867(9?)
155. James Edmund Jones, Pioneer Crimes and Punishment (Toronto, 1924), p. 93
156. *History of Toronto and County of York, Ontario*, Vol. II (Toronto, 1885), p. 150
157. "Autobiography of Philip DeGrassi," Todmorden Mills Museum Archives
158. W. Stewart Wallace, "The Story of Charlotte and Cornelia DeGrassi," *Royal Society of Canada Proceedings and Transactions*, 1941
159. "Autobiography of Philip DeGrassi"
160. "Autobiography of Philip DeGrassi"
161. "Autobiography of Philip DeGrassi"
162. "Diary of William Helliwell," December 5, 1837, Todmorden Mills Museum Archives
163. "Diary of William Helliwell," Dec. 7, 1837
164. James Helliwell, ed., "Diary of William Helliwell"
165. "Diary of William Helliwell," Dec. 10, 1837
166. "William Helliwell Diary," Feb. 24, 1838
167. "Diary of Sarah Glassco Helliwell," Todmorden Mills Museum Archives
168. Kyte, p. 234-5

INDEX

SUMMARY OF OWNERSHIP OF THE DON MILLS
1795-1822

Dates	Owners/Grantees	Operator(s)	Details
1795-June 1797	Isaiah Skinner (1/2 share) Aaron Skinner (1/2 share)	same	Simcoe granted the mill-site to the Skinners on the condition that they build a sawmill at their own expense.
June 1797-1798	Isaiah Skinner	Isaiah Skinner Parshall Terry	Aaron moved back to Niagara around June 1797 and disposed of his share of the property to his brother, Isaiah. Terry moved to the area around July 1797 and entered into some form of partnership with Isaiah concerning the mills in 1798.
1799-1801	Isaiah Skinner (2/3 share) Parshall Terry (1/3 share)	Parshall Terry, assisted by his sons-in-law	Isaiah returned to Niagara in 1799. He sold a 1/3 share of the property to Terry who also leased the balance of the operation.
1801-1805	Timothy Skinner, Sr. (2/3 share) Parshall Terry (1/3 share)	same as above	Isaiah transferred his shares to his father, Timothy Skinner, Sr.
1805-1806	Timothy Skinner, Sr. (2/3 share) Isaiah Skinner (1/3 share)	Parshall Terry, assisted by Samuel Sinclair and the Kendrick brothers	Terry encountered financial difficulties and sold his share of the property back to Isaiah Skinner, but maintained his lease of its operation.
1807-1808	Timothy Skinner, Jr. (2/3 share) Isaiah Skinner (1/3 share)	same as above, although Timothy Jr. was present and may have assisted.	Timothy Skinner Sr. gave his 2/3 share of the property to his younger son, Timothy Jr. Terry continued to hold the lease until his death in July 1808.
August 1808-1809	same as above	unknown others, on behalf of Rhoda Terry, widow of Parshall	Terry drowned in July 1808. His widow surrendered the lease to her father, Timothy Sr. Timothy Jr. returned his 2/3 share of the property to him also.

1809-1811	Timothy Skinner, Sr. (2/3 share) Isaiah Skinner (1/3 share)	Timothy Skinner, Jr.	Timothy Jr. operated the mills on behalf of his father and brother.
1811	Timothy Skinner, Sr.	same as above	Isaiah sold his 1/3 share of the property to his father, Timothy Skinner Sr.
1811-1812	Timothy Skinner, Jr. (2/3) Timothy Skinner, Sr. (1/3)	same as above	Timothy Sr. gave 2/3 share of the property to Timothy Jr.
1812-1814	same as above	Timothy Skinner Jr. when on leave, Samuel Sinclair in his absence.	Timothy Jr. joins the Lincoln Militia and serves during the War of 1812. He was granted at least two leaves to tend to the mills. In his absence, the mills are operated by Samuel Sinclair. Timothy Jr. missing in action and presumed killed at Battle of Chippewa, 1814.
1814-1815	Ann Skinner, widow (2/3 share) Timothy Skinner, Sr. (1/3 share)	Samuel Sinclair, assisted by George Casner	Ann moved to York to await legal recognition of her husband's death. Sinclair leased the mill and operated it with assistance from George Casner. Timothy Skinner Sr. died in 1815.
1815-1817	Ann Skinner, later Ann Sinclair (2/3) Colin Skinner (1/3 share)	same as above	Sinclair continued to operate mills. Married Ann in 1817. Will of Timothy Sr. left 1/3 share of the property to his grandson, Colin Skinner (son of Isaiah Skinner, one of its original owners.)
1817-1820	Henry Skinner (2/3 share) Colin Skinner (1/3 share)	same as above	Samuel Sinclair appointed guardian to Ann's children after their marriage. They sell the property to Henry Skinner, the eldest surviving son of Timothy Skinner, Sr. Joseph Skinner, the son of Timothy Jr., challenges this sale thirty years later, without success.
1820	Colin Skinner	Colin Skinner	Henry Skinner sold his share of the property to Colin. Sinclair left the mills. Colin divided the property and sold part of it to the partnership of Eastwood and Helliwell.

SUMMARY OF OWNERSHIP OF THE PAPER-MILL PORTION OF TODMORDEN MILLS 1822-1855

Dates	Owners	Operator(s)	Details
1822 - 1841	Colin Skinner and John Eastwood	same	Partnership of Thomas Helliwell Sr. and John Eastwood ended in 1822. Eastwood and Skinner formed new partnership and added a distillery to the mills. Opened paper mill in 1826.
1841 - 1850	John Eastwood	John Eastwood	Colin Skinner died in 1841. His share of property sold that year to John Eastwood.
1850 - 1852	Estate of John Eastwood	Colin Skinner Eastwood, as executor of his father's estate	Eastwood died in 1850.
1852 - 1855	Estate of John Eastwood but with claim challenged by Joseph Skinner and his son, Rufus Skinner	unclear. Joseph Skinner may have operated them during some of this time. Mills may also have been closed for some of these years.	Joseph Skinner, son of Timothy Skinner, Jr., challenged the Eastwood ownership of the property, claiming that it was sold illegally in 1817 to Henry Skinner while Joseph was still a minor.
1855	Taylor family	same	Eastwood family reached an agreement with Joseph Skinner and the property was sold to the Taylor family.

SUMMARY OF OWNERSHIP OF THE BREWERY PORTION OF TODMORDEN MILLS
1820 - 1855

Date(s)	Owner(s)	Operator(s)	Details
November 1820 -1822	Thomas Helliwell Sr. and John Eastwood	same	Partnership of Eastwood and Helliwell bought a portion of mill property from Colin Skinner. Built a brewery, malt house and distillery in 1821.
1822 - 1823	Thomas Helliwell and sons	same	Partnership with Eastwood is dissolved. Helliwells retained brewery and distillery business.
1823 - 1842	Sarah Lord Helliwell and sons	same	Thomas Helliwell Sr. died. Left business to wife and sons, each son to become a partner on his 21st birthday.
1842 - 1847	Helliwell brothers	same	Sarah Lord Helliwell died in 1842. Brewery burned in 1847. Partnership dissolved, property divided.
1847 - 1855	Joseph Helliwell and Thomas Helliwell, Jr.	Joseph Helliwell	Joseph built a grist-mill and farmed the property. Property is slowly sold to Taylor family.
1855	Taylor family	same	Virtually complete control of property acquired by Taylor family.

The Mill Families (Skinner)
(names associated with the Don Mills are shown in bold)

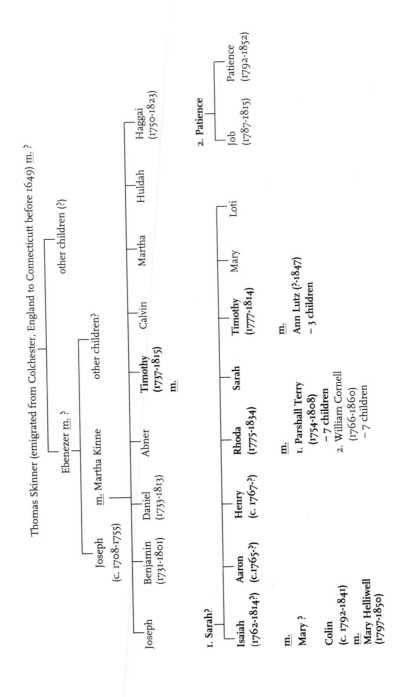

The Mill Families (Helliwell)
(names associated with the Don Mills are shown in bold)

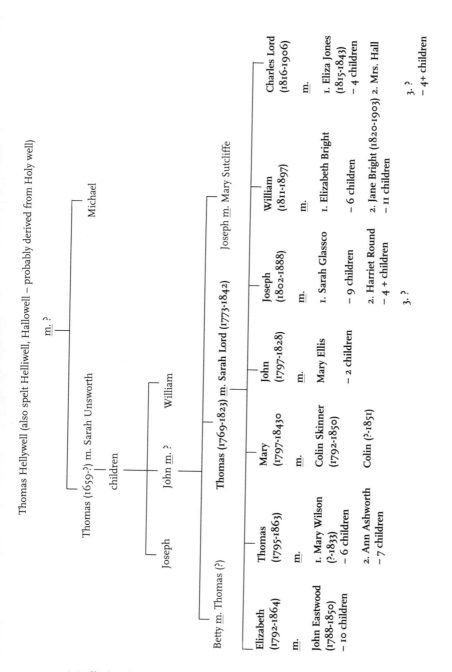

The Mill Families (Terry)
(names associated with the Don Mills are shown in bold)

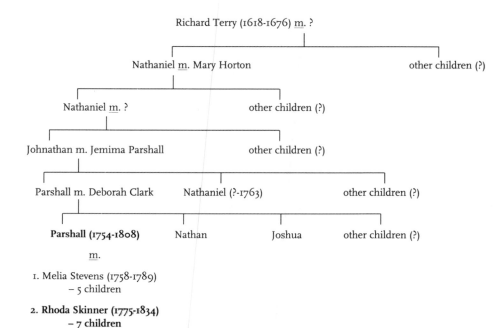

Richard Terry (1618-1676) m. ?

Nathaniel m. Mary Horton

other children (?)

Nathaniel m. ?

other children (?)

Johnathan m. Jemima Parshall

other children (?)

Parshall m. Deborah Clark   Nathaniel (?-1763)   other children (?)

Parshall (1754-1808)   Nathan   Joshua   other children (?)

m.

1. Melia Stevens (1758-1789)
   – 5 children

2. **Rhoda Skinner (1775-1834)**
   – 7 children

BIBLIOGRAPHY OF PUBLISHED SOURCES

Andre, Joseph. *Infant Toronto*. Toronto, 1971.

Baines Yorkshire Directory. London, 1822.

Blake, V.B. *Don Valley Conservation Report*. Toronto, 1950.

Burbank, James W. *Cushetunk: 1754-1784*. Callicoon, N.Y., 1952.

Canadian Almanac and Royal Calendar for 1839. Toronto, 1839.

Carroll, Rev. John. *My Boy Life*. Toronto, 1882.

Carruthers, G. *Paper in the Making*. Toronto, 1947.

Commemorative Biographical Record of the County of York, Ontario. Toronto, 1907.

Cruikshank, E.A., ed. *Ten Years of the Colony of Niagara, 1780-90*. Welland, 1908.

—-. *Simcoe Papers*, Vol. I & V. Toronto, 1923.

—-. *Russell Papers*, Vol. I. Toronto, 1932.

Firth, Edith. *The Town of York, 1793-1815*. Toronto, 1962.

—-. *The Town of York, 1815-1834*. Toronto, 1966.

Fothergill, Charles. *A Few Notes of a Journey from Montreal to Upper Canada*. York, 1817.

Guillet, E.C. *Toronto*. Toronto, 1934.

Helliwell, Albert F. *Helliwell Family Record*. Portland, 1959.

History of Toronto and County of York, Ontario, Vol. II, Toronto, 1885.

Jones, James Edmund. *Pioneer Crimes and Punishments in Toronto and the Home District*. Toronto, 1924.

Kyte, E.C. *Old Toronto*. Toronto, 1954.

Lund, Norah Hall. *Parshall Terry Family History*. Paragonah, 1956.

Macpherson, K.R. *The King's Mill, A Resume*. Toronto, 1963.

Middleton, J.E. *The Municipality of Toronto, Vol. II*. Toronto, 1923.

——. "Diary of Joseph Willcocks," *History of Ontario*, Toronto, 1927.

Moody, James. *Lt. James Moddy's Narrative of his Exertions and Sufferings*, reprinted. New York, 1968.

Newell, Diane. *Dictionary of Canadian Biography, Vol. IX, 1861-70*. Toronto, 1976.

Robertson, John Ross. *Landmarks of Toronto, Vol. VI*. Toronto, 1894.

——. *History of Freemasonry in Canada*. Toronto, 1899.

——. *The Diary of Mrs. John Graves Simcoe*. Toronto, 1911.

Robinson, Percy. *Toronto during the French Regime*. Toronto, 1933.

Sabine, Lorenzo. *American Loyalists or Biographical Sketches of Adherents of the British Crown*. Boston, 1847.

Scadding, Henry. *Toronto of Old*. Toronto, 1873.

Smith, W.H. *Canada Past, Present and Future*. Toronto, 1851.

Strachan, James. *A Visit to the Province of Upper Canada in 1819*, reprinted. Toronto, 1968.

Ure, George. *Handbook of Toronto*. Toronto, 1858.

Williamson, Joseph. *History of Belfast, Maine*. Portland, 1877.

ELEANOR DARKE

Eleanor Darke was raised in the Toronto suburb of "Don Mills", little guessing that she would later begin her career at the museum which preserved the remains of the village from which it derived its name. After graduating with an Honours B.A. in History from York University in 1973, she took up the post of curator-Manager at the Todmorden Mills Museum. In 1984 she was hired by the Toronto Historical Board to manage the Mackenzie House Museum. She has recently been seconded to the Toronto Historical Board's headquarters at 205 Yonge Street to co-ordinate its walking tours and lectures and to assist with special programmes.

IAN WHEAL

The mill-sites early history was extensively researched by Ian Wheal over a four year period (1977-1981). Starting out as the historical researcher in the summer of 1977, he continued on with the work as a museum volunteer.

Since then he has been involved in historical projects throughout Ontario, most notably at Fort Frances, Kingston and Missinaibi Provincial Park near Chapleau.

In recent years he has been actively involved in Ontario's Industrial Heritage. He is a former editor of the *Ontario Society for Industrial Archeology Newsletter* and is presently a director of that society.

Most recently, he was Project Co-ordinator and Historical Researcher for the Exhibition: A Hazardous Crossing, Toronto's Great Railway Viaduct, 1925-1930, at the Market Gallery, Toronto.